Maria Moszyńska

Selected Topics in
Convex Geometry

Birkhäuser
Boston • Basel • Berlin

Maria Moszyńska
Warsaw University
Institute of Mathematics
02-097 Warsaw
Poland

Cover design by Joseph Sherman.

Mathematics Subject Classification (2000): 52-01, 52-02, 52A20, 52A30, 52A35, 52A38, 52A40, 52B11, 52B45

ISBN-10 0-8176-4396-6   eISBN 0-8176-4451-2   Printed on acid-free paper.
ISBN-13 978-0-8176-4396-6

©2006 Birkhäuser Boston   *Birkhäuser*
Based on the original Polish edition, *Geometria zbiorów wypukłych. Zagadnienia wybrane*,
Wydawnictwa Naukowo-Techniczne, Warszawa, Poland, ©2001.

Printed in the United States of America.   (TXQ/MP)

9 8 7 6 5 4 3 2 1

www.birkhauser.com

Contents

Part III

Preface to the Polish Edition

This book (more precisely, its Part I and the beginning of Part II) is based on the monographic lecture under a similar title, presented several times, in various versions, at the Department of Mathematics, Computer Science, and Mechanics of Warsaw University. I hope it will be helpful for lectures and seminars. Perhaps it may be of interest for mathematicians working in geometry (in a broad sense) or in other fields. I will be happy if the book is useful as well for those using mathematics as a tool.

I shall be grateful to readers for their critical remarks.[1]

Many people deserve my thanks. The first is Professor Jakub Bodziony. It was due to him that I heard of stereology, which, to some extent, is based on the Hadwiger theorems and the Crofton formulae. These famous results are at the heart of Part I of my book. In turn, my contact with Jakub Bodziony was due to my colleagues Andrzej Palczewski and the late Wiesław Szlenk, who in 1989 organized a seminar on mathematical methods in natural science.

It is hard to overestimate the merits of Krzysztof Przesławski, whose remarks made it possible for me to avoid many mistakes and to improve the text.

I wish to thank Tomasz Żukowski and my son, Marcin Moszyński, for their help and advice.

I am grateful to Adam and Agnieszka Bogdewicz for their cooperation in preparing the figures.

Warszawa, April 2001   Maria Moszyńska

---

[1] E-mail: mariamos@mimuw.edu.pl

Preface to the English Edition

Following a suggestion of Peter Gruber, Erwin Lutwak, and Carla Peri, I decided to translate my book into English. My motivation was obvious: very few people interested in the subject can read a Polish book.

Peter Gruber's advice was to start at least one year after the Polish edition appeared, because usually there are many errors to be found and corrected. He was right: in the first year after the appearance of the book (autumn 2001 to autumn 2002), my colleagues, students, and I found many things to be improved. Thus, I used the opportunity to make corrections, to complete or add some proofs, to reorganize parts of the text, and to update some of the results and references.

I very much appreciate remarks and suggestions of the referees.

I am grateful to everybody who found an error in the Polish edition.

My special thanks go to my friend Irmina Herburt; our cooperation has played a very essential role in preparing the new edition.

I hope that this new version, though certainly not perfect, is much better than the old one.

Warsaw, September 2004   Maria Moszyńska

Introduction

Although convex geometry has its roots in the middle of the nineteenth century, only in the last few decades has it become one of the most vivid branches of contemporary geometry.

The book consists of three parts. In Part I, the Hadwiger theorems on functionals are presented as the main topic, with the classical proof, which was the only existing one until 1995 (compare [64] and [38]). Chapters 1–3 concern basic notions of metric geometry, in particular of Euclidean geometry, and geometry of convex sets. Chapter 4 deals with maps of the family $\mathcal{K}^n$ of compact convex subsets of $\mathbf{R}^n$ into itself, for instance the Steiner symmetrization, which is used in Chapter 5 in the proof of one of two rounding theorems (in German *Kugelungstheoreme*, sometimes called also theorems on a snowball). Chapter 6 deals with convex polytopes, their role in the class $\mathcal{K}^n$ (approximation theorems), and the equivalence by dissection. Chapter 7 is devoted to functionals on $\mathcal{K}^n$, in particular basic functionals (*intrinsic volumes*) and the Steiner theorem. The Hadwiger theorems are the subject of Chapter 8. In Chapter 9 (the last one in Part I) one can find their applications, in particular the Crofton formulae, which can be considered as the origin of geometric tomography (see [20]).

Part II (Chapters 10–13) deals with various generalizations of notions previously considered.

First, in Chapter 10, we give a survey (based on [62]) of curvature measures and surface area measures, which are an important tool of convex geometry and may be treated as a "localization" of intrinsic volumes. By the use of curvature measures we show the relationship between the mean curvature considered in Section 9.1 and the integral of the mean curvatures in the sense of differential geometry.

Further, in Chapter 11, various extensions of the class $\mathcal{K}^n$ are presented: the class of *sets with positive reach* and the *convexity ring* $\mathcal{U}^n$ (called *convex ring* in the literature).

In Chapter 12 functionals on $\mathcal{K}^n$ are replaced by selectors, which select a point from every convex body.

Chapter 13 concerns one more operation on the class $\mathcal{K}^n$: polarity.

Part III, the shortest one (Chapters 14–16), is devoted to the class of star sets, which recently turned out to be a useful tool of geometric tomography (see [20], a very good monograph being updated by the author in the Internet).[2] This domain is too rich to be developed in this book, but because of its essential role it cannot be avoided.

The whole book with the exception of Chapter 1 concerns Euclidean spaces. The restriction to $\mathbb{R}^n$ (with the Euclidean metric) is justified by the fact that every n-dimensional Euclidean space is isometric to $\mathbb{R}^n$. We do not follow the tendency to distinguish between the affine space $\mathbb{R}^n$ and the corresponding linear space; thus we identify a point $x = (x_1, \ldots, x_n) \in \mathbb{R}^n$ with the vector with coordinates $x_1, \ldots, x_n$. Consequently, the unit sphere $S^{n-1} := \{x \in \mathbb{R}^n \mid \|x\| = 1\}$, which is the boundary of the unit ball $B^n := \{x \in \mathbb{R}^n \mid \|x\| \leq 1\}$, is often treated as the set of vectors with length equal to 1 (compare [64]).

In principle, we follow the terminology and notation used in [64]. Thus, $\mathcal{B}(X)$ is the class of Borel subsets of X; the function $V_n : \mathcal{B}(\mathbb{R}^n) \to \mathbb{R}$ is the n-*dimensional volume*, i.e., $V_n = \lambda_n$, the Lebesgue measure; σ_{n-1} is the $(n-1)$-dimensional spherical Lebesgue measure; and

$$\kappa_n := V_n(B^n), \quad \omega_n := \sigma_{n-1}(S^{n-1}) = n\kappa_n$$

(compare Proposition 7.3.4). Often, if it does not lead to confusion, we write σ instead of σ_{n-1}.

The notions defined for $\mathbb{R}^k$, $k \leq n$, invariant under isometries (or more generally, under affine maps) can be transferred in the obvious way to an arbitrary k-dimensional affine subspace of $\mathbb{R}^n$.

For any nonempty sets X and X', a function $f : X \to X'$ is called a *surjection* if $f(X) = X'$, an *injection* if different points have different values, and a *bijection* if it is both a surjection and an injection.

For metric spaces (X, ϱ) and (X', ϱ'), a function $f : X \to X'$ is a Lipschitz map if and only if there exists a $\lambda > 0$ satisfying the condition

$$\varrho'(f(a), f(b)) \leq \lambda\varrho(a, b) \quad \text{for every} \quad a, b \in X; \qquad (0.1)$$

the infimum of the set of positive numbers satisfying (0.1) is called the *Lipschitz constant of* f. A function f with the Lipschitz constant $\lambda \leq 1$ is called a *weak contraction*, and a function with the Lipschitz constant $\lambda < 1$ is called a *contraction*. Evidently, every Lipschitz function is uniformly continuous.

---

[2] http://www.ac.wwu.edu/~gardner/

A map $r : X \to r(X) \subset X$ is a *retraction of a space* (X, ϱ) provided that $r|r(X) = \mathrm{id}_{r(X)}$. A subset X_0 of the space X is a *retract of* X if there exists a retraction $r : X \to X_0$ (compare [12]).

Let us notice that there is a difference between our terminology concerning isometries of arbitrary metric spaces and the terminology used by P. Gruber ([25]). We require an *isometry* to be a surjection, while any map preserving distances is referred to as *isometric embedding* (Definition 1.1.10, Theorem 1.1.11, Example 4.1.2).

The symbol card denotes cardinality. The symbol $\mathbf{P}_{k=1}^{\infty}$ denotes the countable Cartesian product of a sequence of sets:

$$\mathbf{P}_{k=1}^{\infty} X_k := \{(x_k)_{k\in\mathbb{N}} \mid x_k \in X_k \quad \text{for every } k\}.$$

The symbols cl_ϱ, int_ϱ, bd_ϱ denote, respectively, *closure, interior, boundary in the space* (X, ϱ). Usually we omit the subscript ϱ if it does not lead to confusion. We use the notation

$$\mathrm{dist}(A, B) := \inf\{\varrho(x, y) \mid x \in A, \ y \in B\}. \tag{0.2}$$

The symbols lin, aff, and pos are used for subsets of $\mathbb{R}^n$: $\mathrm{lin}A$ is the smallest linear subspace containing A, $\mathrm{aff}A$ the smallest affine subspace containing A; for every $x \in \mathbb{R}^n \setminus \{0\}$,

$$\mathrm{pos}x := \{\lambda x \mid \lambda \geq 0\},$$

and for every $X \subset \mathbb{R}^n$,

$$\mathrm{pos}X := \bigcup\{\mathrm{pos}x \mid X \cap \mathrm{pos}x \neq \emptyset\}.$$

The symbols $\mathrm{relint}A$ and $\mathrm{relbd}A$ denote, respectively, the *relative interior* and the *relative boundary* of a subset A of $\mathbb{R}^n$, i.e., the interior or the boundary of A with respect to $\mathrm{aff}A$.

Further, $\mathcal{G}_k^n$ is the set of k-dimensional linear subspaces of $\mathbb{R}^n$; for every $E \in \mathcal{G}_k^n$, the set $E^\perp$ is the $(n - k)$-dimensional linear subspace orthogonal to E; for every nonzero vector v we admit $v^\perp := (\mathrm{lin}v)^\perp$. For every k-dimensional affine subspace E in $\mathbb{R}^n$ and $x \in \mathbb{R}^n$, the set $E^\perp(x)$ is the $(n - k)$-dimensional affine subspace passing through x and perpendicular to E; the function $\pi_E : \mathbb{R}^n \to E$ is the orthogonal projection onto E, and σ_E is the symmetry with respect to E. The usual scalar product in $\mathbb{R}^n$ is denoted by $\circ$.

$GL(n)$, $O(n)$, and $SO(n)$ are, respectively, the group of linear automorphisms, the group of linear isometries, and the group of proper linear isometries of $\mathbb{R}^n$; finally, Tr is the group of translations.

The symbol $\equiv_G$ denotes the congruence of sets with respect to a group G of transformations.

The symbol $\overset{w}{\to}$ denotes the weak convergence of measures on a metric space, here on $\mathbb{R}^n$:

$$\mu_i \overset{w}{\to} \mu :\Longleftrightarrow \forall X \in \mathcal{B}(\mathbb{R}^n)\, (\mu(\mathrm{bd}X) = 0 \Longrightarrow \lim_{i \to \infty} \mu_i(X) = \mu(X))$$

(compare [4]).

$\mathcal{H}^m$ is the m-dimensional Hausdorff measure (see [16]), which for subsets of $\mathbb{R}^m$ coincides with the m-dimensional Lebesgue measure λ_m, and for subsets of S^m with the spherical measure σ_m (see [64]).

The symbol δ_i^j denotes 0 or 1, respectively, for $i \neq j$ or $i = j$.

We use quantifiers: $\forall$ (for every), $\exists$ (there exists), and $\exists^1$ (there exists exactly one). In definitions the word "if" is always understood as "if and only if."

At the end of the book the reader will find exercises. Some of them (but not all) are quite elementary.

Part I

1

Metric Spaces

Let (X, ϱ) be a metric space. Thus, X is a nonempty set and the function $\varrho :$ $X \times X \to R_+$, referred to as a *metric*, satisfies the conditions

($\star$) $\varrho(x, y) = 0$ if and only if $x = y$,

($\star\star$) $\varrho(x, y) = \varrho(y, x)$,

($\star\star\star$) $\varrho(x, y) + \varrho(y, z) \geq \varrho(x, z)$.

(We omitted the universal quantifiers at the beginning of these sentences. In what follows, we shall often do this when it does not lead to confusion.)

1.1 Distance of point and set. Generalized balls

1.1.1. DEFINITION. For every nonempty subset A of X and $x \in X$ let

$$\varrho(x, A) := \inf\{\varrho(x, a) \mid a \in A\}.$$

The number $\varrho(x, A)$ is the *distance between the point x and the set A*.

Let us notice that

1.1.2. *The function $\varrho(\cdot, A) : X \to R_+$ is continuous.*

Proof. Let $x = \lim x_k$. By the triangle inequality (condition ($\star\star\star$) above) and properties of upper bound,

$$-\varrho(x_k, x) + \varrho(x, A) \leq \varrho(x_k, A) \leq \varrho(x_k, x) + \varrho(x, A).$$

Thus $\varrho(x, A) = \lim \varrho(x_k, A)$. □

Moreover, the function $\varrho(\cdot, A)$ is a weak contraction (Exercise 1.1).

1.1.3. $\varrho(x, A) = \varrho(x, \mathrm{cl}A)$.

Proof. Since $A \subset \mathrm{cl}A$, it follows that

$$\varrho(x, A) \geq \varrho(x, \mathrm{cl}A).$$

Hence, it suffices to prove that for every $a \in \mathrm{cl}A$,

$$\varrho(x, A) \leq \varrho(x, a). \tag{1.1}$$

If $a \in \mathrm{cl}A$, then $a = \lim a_k$ for some sequence $(a_k)_{k \in \mathbb{N}}$ in A; thus $\varrho(x, A) \leq \varrho(x, a_k)$ for every k. Passing to the limit for $k \to \infty$, by 1.1.2 (for a singleton) we obtain (1.1). □

Using 1.1.2 and 1.1.3, it is easy to prove

1.1.4. $\varrho(x, A) = 0 \quad \Leftrightarrow \quad x \in \mathrm{cl}A$.

1.1.5. DEFINITION. For every $A \subset X$ and $\varepsilon > 0$, let

$$(A)_\varepsilon := \{x \in X \mid \varrho(x, A) \leq \varepsilon\}.$$

The set $(A)_\varepsilon$ is called *ε-hull of A* or *generalized ball of A*.

Evidently,

1.1.6. *For every nonempty $A, B \subset X$ and $\delta, \varepsilon > 0$,*

$$A \subset B \Longrightarrow (A)_\varepsilon \subset (B)_\varepsilon$$

and

$$((A)_\delta)_\varepsilon \subset (A)_{\delta+\varepsilon}.$$

(Compare Exercise 1.7.)

It is also easy to prove that

1.1.7. *If A is compact, then*

$$(A)_\varepsilon = \bigcup_{a \in A} \{a\}_\varepsilon.$$

(Compactness is essential for the inclusion $\subset$, while $\supset$ is true for arbitrary A.)

The set $\{a\}_\varepsilon$ is the *ball with center a and radius ε*; it is denoted by the symbol $B(a, \varepsilon)$ (or $B_X(a, \varepsilon)$ if it is not obvious that the ball is taken in (X, ϱ)).

1.1.8. DEFINITION. The set $A \subset X$ is *bounded* if there exists an upper bound of the set $\{\varrho(x, y) \mid x, y \in A\}$. This upper bound is called the *diameter of A* (in symbols diam$A$.

It is easy to check that

1.1.9. *For every $A \subset X$ the following conditions are equivalent:*
(i) *A is bounded,*
(ii) $\exists \varepsilon > 0 \; \exists x \in X \; A \subset B_X(x, \varepsilon)$.

1.1.10. DEFINITION. For arbitrary metric spaces (X, ϱ) and (X', ϱ'), a function $f : X \to X'$ is *an isometric embedding (with respect to ϱ and ϱ')* if

$$\forall x, y \in X \; \varrho'(f(x), f(y)) = \varrho(x, y); \tag{1.2}$$

a surjective isometric embedding is called an *isometry*. More generally, a surjection f is a *similarity with ratio $\lambda > 0$* if

$$\forall x, y \in X \; \varrho'(f(x), f(y)) = \lambda \varrho(x, y).$$

The spaces (X, ϱ) and (X', ϱ') are *isometric (similar)* if there exists an isometry (a similarity) $f : X \to X'$.

It is well known that $\mathbf{R}^n$ is not isometric with any of its proper subsets. We give here a complete proof of this assertion.[1]

1.1.11. THEOREM. *Every isometric embedding $f : \mathbf{R}^n \to \mathbf{R}^n$ is an isometry.*

Proof. Let $X = f(\mathbf{R}^n)$. Of course, f treated as a function of $\mathbf{R}^n$ onto X is an isometry, whence it preserves completeness and connectedness; thus X is a closed connected subset of $\mathbf{R}^n$. Obviously, X is not compact, since f is a homeomorphism of $\mathbf{R}^n$ on X.

Let $n = 1$. Suppose that $X \neq \mathbf{R}$. Then X is a closed half-line with an endpoint a. The point a does not disconnect X (i.e., $X \setminus \{a\}$ is connected), while every point of $\mathbf{R}$, in particular $f^{-1}(a)$, disconnects $\mathbf{R}$. But this is impossible because f is a homeomorphism.

Now let $n \geq 2$. Let us notice that the image $f(L)$ of any line $L \subset \mathbf{R}^n$ is again a line. Indeed, let $\phi : \mathbf{R} \to L$ be an isometry of $\mathbf{R}$ on L; then the function $\psi : \mathbf{R} \to f(L)$ defined by

$$\psi(x) := f\phi(x)$$

is an isometry of $\mathbf{R}$ on $f(L)$.

For any $p \in X$, the set $\mathbf{R}^n$ is a union of the family $\mathcal{L}$ of all the lines passing through p:

$$\mathbf{R}^n = \bigcup \mathcal{L}. \tag{1.3}$$

For every $L \in \mathcal{L}$,

$$f^{-1}(X \cap L) = f^{-1}(X) \cap f^{-1}(L) = \mathbf{R}^n \cap f^{-1}(L) = f^{-1}(L),$$

whence $X \cap L = L$; thus $L \subset X$. Consequently, $\mathbf{R}^n \subset X$, by (1.3). Hence $X = \mathbf{R}^n$. □

As we shall see in Chapter 4, Theorem 1.1.11 cannot be generalized on arbitrary metric spaces (Example 4.1.2).

---

[1] It is a consequence of theorem on perfect metric homogeneity of $\mathbf{R}^n$ [11].

1.2 The Hausdorff metric

Let $\mathcal{C}(X)$ be the class of nonempty, closed, bounded subsets of a metric space (X, ϱ). For any $A, B \in \mathcal{C}(X)$, let

$$\varrho_H(A, B) := \inf\{\varepsilon > 0 \mid A \subset (B)_\varepsilon \text{ and } B \subset (A)_\varepsilon\}. \qquad (1.4)$$

(The lower bound exists, since the sets A and B are bounded.)

We shall prove that

1.2.1. *The function* $\varrho_H : \mathcal{C}(X) \times \mathcal{C}(X) \to \mathbb{R}$ *is a metric.*

Proof. Obviously, $\varrho_H \geq 0$. Let $A, B \in \mathcal{C}(X)$. Since A and B are closed in X, by 1.1.4 it follows that

$$\bigcap_{\varepsilon > 0}(A)_\varepsilon = A \text{ and } \bigcap_{\varepsilon > 0}(B)_\varepsilon = B.$$

Hence

$$\varrho_H(A, B) = 0 \Leftrightarrow \forall \varepsilon > 0 \, (A \subset (B)_\varepsilon \text{ and } B \subset (A)_\varepsilon)$$

$$\Leftrightarrow A \subset B \text{ and } B \subset A \Leftrightarrow A = B.$$

This proves condition $(\star)$.

Evidently ϱ_H satisfies $(\star \star)$. It remains to verify $(\star \star \star)$.

Let $A, B, C \in \mathcal{C}(X)$; in view of $\star$, we may assume that A, B, C are pairwise distinct. Let $\varepsilon_0 := \varrho_H(A, B)$ and $\delta_0 := \varrho_H(B, C)$.

It is easy to see that the set $\{\varepsilon > 0 \mid A \subset (B)_\varepsilon \text{ and } B \subset (A)_\varepsilon\}$ is closed, whence its lower bound $\varrho_H(A, B)$ belongs to it, i.e.,

$$A \subset (B)_{\varepsilon_0} \text{ and } B \subset (A)_{\varepsilon_0}.$$

Similarly,

$$B \subset (C)_{\delta_0} \text{ and } C \subset (B)_{\delta_0}.$$

Hence in view of 1.1.6, $A \subset (C)_{\varepsilon_0 + \delta_0}$ and $C \subset (A)_{\varepsilon_0 + \delta_0}$. Therefore,

$$\varrho_H(A, C) \leq \varepsilon_0 + \delta_0 = \varrho_H(A, B) + \varrho_H(B, C). \qquad \square$$

The metric ϱ_H is called the *Hausdorff metric*; the limit in the space $(\mathcal{C}(X), \varrho_H)$ is called the *Hausdorff limit*:

$$A = \lim_H A_n \Leftrightarrow \lim \varrho_H(A, A_n) = 0.$$

The following formula (1.5) is often given as a definition of the Hausdorff metric.

1.2.2. THEOREM. *For every $A, B \in C(X)$,*

$$\varrho_H(A, B) = \max\{\sup_{a \in A} \varrho(a, B), \sup_{b \in B} \varrho(b, A)\}. \qquad (1.5)$$

Proof. Since for every connected $S_1, S_2 \subset R_+$ with nonempty intersection

$$\inf(S_1 \cap S_2) = \max\{\inf S_1, \inf S_2\},$$

by the symmetry of condition (1.5) with respect to A and B it suffices to prove that

$$\sup_{a \in A} \varrho(a, B) = \inf\{\varepsilon > 0 \mid A \subset (B)_\varepsilon\}.$$

Let

$$\alpha := \sup_{a \in A} \varrho(a, B) \quad \text{and} \quad \beta := \inf\{\varepsilon > 0 \mid A \subset (B)_\varepsilon\}.$$

Then $\varrho(a, B) \leq \alpha$ for every $a \in A$, and thus $A \subset (B)_\alpha$; hence $\alpha \geq \beta$.

Suppose $\alpha > \beta$; then

$$\exists \varepsilon \in (0; \alpha) \ A \subset (B)_\varepsilon,$$

whence $\sup_{a \in A} \varrho(a, B) \leq \varepsilon < \alpha$, contrary to the assumption. $\qquad \square$

The properties of the space $(C(X), \varrho_H)$ obviously depend on those of (X, ϱ) (see Theorem 1.2.6).

1.2.3. DEFINITION. A space (X, ϱ) is *finitely compact* if every closed, bounded subset of (X, ϱ) is compact.

Let us note

1.2.4. PROPOSITION. *For every metric space (X, ϱ) the following conditions are equivalent:*
   (i) *(X, ϱ) is finitely compact;*
   (ii) *every ball in (X, ϱ) is compact;*
   (iii) *every bounded sequence in (X, ϱ) has a convergent subsequence.*

Evidently, every compact space is finitely compact. The space R^n is finitely compact but not compact. This example might suggest that completeness implies finite compactness. But this implication is false; for instance, the plane R^2 with the "railway metric" $\tilde{\varrho}$ defined by

$$\tilde{\varrho}(x, y) = \begin{cases} \|x - y\| & \text{if } 0 \in \text{aff}(x, y) \\ \|x\| + \|y\| & \text{if } 0 \notin \text{aff}(x, y) \end{cases}$$

is complete but is not finitely compact, since the balls with center $(0, 0)$ are not compact. Similarly, the space l^2, i.e., the space of real sequences with convergent series of squares, with the metric ϱ defined by the formula

$$\varrho((x_i)_{i \in N}, (y_i)_{i \in N}) = (\sum_{i=1}^{\infty} (x_i - y_i)^2)^{\frac{1}{2}},$$

is complete but is not finitely compact.

Since the balls in every metric space are closed and bounded, it follows that every finitely compact space is locally compact.

We shall need the following

1.2.5. LEMMA. *If a space (X, ϱ) is finitely compact, then for every descending sequence $(A_n)_{n \in \mathbb{N}}$ in $\mathcal{C}(X)$,*

$$\bigcap_{n=1}^{\infty} A_n = \lim_{\mathrm{H}} A_n.$$

Proof. Let $A = \bigcap_{n=1}^{\infty} A_n$. In view of the finite compactness of (X, ϱ), by the Cantor theorem, the set A is nonempty. Since $A \subset A_n$ for every n, it follows that moreover,

$$\forall \varepsilon > 0 \; \forall n \; A \subset (A_n)_\varepsilon.$$

It remains to prove that

$$\forall \varepsilon > 0 \; \exists n_0 \; \forall n > n_0 \; A_n \subset (A)_\varepsilon.$$

Suppose, to the contrary, that there exist $\varepsilon > 0$ and an increasing sequence $(k_n)_{n \in \mathbb{N}}$ such that

$$A_{k_n} \not\subset (A)_\varepsilon. \tag{1.6}$$

Let $X_n := A_{k_n} \setminus \mathrm{int}(A)_\varepsilon$ for every n and $X_0 := \bigcap_{n=1}^{\infty} X_n$. Evidently, (X_n) is a descending sequence of compact sets; by (1.6), they are nonempty. Hence by the Cantor theorem,

$$X_0 \neq \emptyset. \tag{1.7}$$

On the other hand, $X_0 \cap A = A \setminus \mathrm{int}(A)_\varepsilon = \emptyset$ and $X_0 \subset \bigcap_{n=1}^{\infty} A_{k_n} = A$ (the last equality holds because the sequence $(k_n)_{n \in \mathbb{N}}$ is increasing), whence $X_0 = X_0 \cap A = \emptyset$, contrary to (1.7). □

We are now ready to prove

1.2.6. THEOREM (compare [64], Th.1.8.2). *If (X, ϱ) is finitely compact, then $(\mathcal{C}(X), \varrho_{\mathrm{H}})$ is complete.*

Proof. Let $(C_n)_{n \in \mathbb{N}}$ be a Cauchy sequence in $(\mathcal{C}(X), \varrho_{\mathrm{H}})$. Then

$$\forall \varepsilon > 0 \; \exists n_0 \; \forall n_1, n_2 \geq n_0 \; C_{n_1} \subset (C_{n_2})_{\frac{\varepsilon}{2}};$$

thus in particular,

$$C_n \subset (C_{n_0})_{\frac{\varepsilon}{2}} \text{ and } C_{n_0} \subset (C_n)_{\frac{\varepsilon}{2}} \text{ for } n > n_0. \tag{1.8}$$

Hence the set $\bigcup_{n=1}^{\infty} C_n$ is bounded, since it is a subset of $(C_{n_0})_\varepsilon \cup \bigcup_{n=1}^{n_0} C_n$.

For every $m \in \mathbb{N}$, let

$$A_m := \mathrm{cl}\left(\bigcup_{n=m}^{\infty} C_n\right) \text{ and } A := \bigcap_{m=1}^{\infty} A_m. \tag{1.9}$$

The set A_m is closed and bounded; thus it is compact, because (X, ϱ) is finitely compact; obviously, $A_{m+1} \subset A_m$ for every $m \in \mathbb{N}$. Hence by Lemma 1.2.5,

$$A = \lim_H A_m,$$

which together with (1.9) implies that there exists n_1 such that

$$C_n \subset \mathrm{cl}\left(\bigcup_{i=n}^{\infty} C_i\right) \subset (A)_\varepsilon \quad \text{for } n > n_1.$$

By (1.8) and (1.9),

$$A \subset \mathrm{cl}\bigcup_{i=n}^{\infty} C_i \subset (C_n)_\varepsilon \quad \text{for } n > n_0.$$

Thus $A = \lim_H C_n$; hence $(C_n)_{n \in \mathbb{N}}$ is convergent. □

1.2.7. COROLLARY. *The space $\mathcal{C}(\mathbb{R}^n)$ is complete.*

Finally, we mention without proof the following

1.2.8. THEOREM (compare [64], 1.8.3 or 1.8.4). *The space $\mathcal{C}(\mathbb{R}^n)$ is finitely compact.*

Let us observe that Theorem 1.2.8 is stronger than 1.2.7 (compare with Exercise 1.5).

2

Subsets of Euclidean Space

In what follows we are concerned with subsets of R^n for $n \geq 1$, or more precisely, of the metric space (R^n, ϱ), where ϱ is the *Euclidean metric*, i.e., the metric induced by the *Euclidean norm*:

$$\varrho(x, y) := \|x - y\| = \sqrt{\sum_{i=1}^{n}(x_i - y_i)^2}$$

for $x = (x_1, \ldots, x_n)$ and $y = (y_1, \ldots, y_n)$.

We use the notation

$$\mathcal{C}^n := \mathcal{C}(\mathrm{R}^n).$$

Since R^n is finitely compact, it follows that $\mathcal{C}^n$ is the class of compact, nonempty subsets of R^n.

2.1 The Minkowski operations

For subsets of R^n, addition and multiplication by a scalar are defined:

2.1.1. DEFINITION. (i) For any $A, B \subset \mathrm{R}^n$,

$$A + B := \{a + b \mid a \in A, \ b \in B\}.$$

The set $A + B$ is the *Minkowski sum* of A and B.

(ii) For any $A \subset \mathbf{R}^n$ and $t \in \mathbf{R}$,

$$tA := \{ta \mid a \in A\}.$$

The set tA is the *product of A by t*.

It is clear that tA is the image of A under the homothety with center 0 and ratio t.

As direct consequences of 2.1.1, we obtain the following two simple statements.

2.1.2. (i) *The singleton {0} is the neutral element of Minkowski addition;*
(ii) *addition is associative and commutative;*
(iii) *multiplication by a scalar is distributive with respect to addition.*

2.1.3. *Minkowski addition and multiplication by a scalar preserve inclusion:*

$$A_i \subset B_i \text{ for } i = 1, 2 \implies A_1 + A_2 \subset B_1 + B_2,$$

$$A \subset B \implies tA \subset tB.$$

It is easy to see that

2.1.4. *The Minkowski operations preserve compactness.*

Hence $+$ is a function from $\mathcal{C}^n \times \mathcal{C}^n$ to $\mathcal{C}^n$, and multiplication by t is a function from $\mathcal{C}^n$ to $\mathcal{C}^n$.

Adding a ball with radius ε to a compact set, we obtain its ε-hull:

2.1.5. *For any $A \in \mathcal{C}^n$ and $\varepsilon > 0$,*

$$(A)_\varepsilon = A + \varepsilon B^n.$$

Proof. Let us fix an $x \in \mathbf{R}^n$. Since the metric, and so also its restriction $\varrho|\{x\} \times A$, is continuous, by the compactness of A it follows that

$$x \in (A)_\varepsilon \iff \exists a \in A \; \|x - a\| \leq \varepsilon \iff x \in A + \varepsilon B^n. \qquad \square$$

The following proposition, which describes the Minkowski operations on generalized balls, is based on 2.1.2 and 2.1.5.

2.1.6. PROPOSITION. (i) *For any $A, B \in \mathcal{C}^n$ and $\alpha, \beta > 0$,*

$$(A)_\alpha + (B)_\beta = (A + B)_{\alpha+\beta};$$

(ii) *for any $A \in \mathcal{C}^n$, $t > 0$, and $\varepsilon > 0$,*

$$t(A)_\varepsilon = (tA)_{t\varepsilon}.$$

Proof. (i): Obviously,

$$\alpha B^n + \beta B^n = (\alpha + \beta) B^n. \tag{2.1}$$

Thus

$$(A)_\alpha + (B)_\beta = A + B + (\alpha + \beta) B^n = (A + B)_{\alpha+\beta}.$$

(ii):

$$t(A)_\varepsilon = t(A + \varepsilon B^n) = tA + t\varepsilon B^n = (tA)_{t\varepsilon}. \qquad \square$$

We shall now prove

2.1.7. THEOREM. (i) *Minkowski addition is continuous on $C^n \times C^n$.*
(ii) *Multiplication by a nonnegative scalar is continuous on C^n.*

Proof. (i): As is well known, convergence in a Cartesian product is equivalent to convergence "by coordinates" independently of the choice of a product metric[1].[2] Hence, it suffices to prove that

$$\forall \varepsilon > 0 \; \exists \delta > 0 \; \forall A_1, A_2, B_1, B_2 \in C^n \quad \varrho_H(A_i, B_i) \leq \delta \text{ for } i = 1, 2$$

$$\implies \varrho_H(A_1 + A_2, B_1 + B_2) \leq \varepsilon. \tag{2.2}$$

Let $\varepsilon > 0$ and let $\delta := \frac{\varepsilon}{2}$. If $\varrho_H(A_i, B_i) \leq \delta$ for $i = 1, 2$, then

$$A_i \subset (B_i)_\delta \text{ and } B_i \subset (A_i)_\delta,$$

whence by 2.1.3 and 2.1.6(i),

$$A_1 + A_2 \subset (B_1 + B_2)_\varepsilon \text{ and } B_1 + B_2 \subset (A_1 + A_2)_\varepsilon,$$

i.e.,

$$\varrho_H(A_1 + A_2, B_1 + B_2) \leq \varepsilon.$$

(ii): If $t = 0$, then $tA = \{0\}$ for every $A \in C^n$; thus the continuity of multiplication by 0 is evident.

Let $t > 0$. For any $\varepsilon > 0$ let $\delta := \frac{\varepsilon}{t}$. If $\varrho_H(A, B) \leq \delta$, then $A \subset (B)_\delta$ and $B \subset (A)_\delta$, whence by 2.1.3 and 2.1.6(ii),

$$tA \subset (tB)_\varepsilon \text{ and } tB \subset (tA)_\varepsilon.$$

Thus

$$\varrho_H(tA, tB) \leq \varepsilon. \qquad \square$$

2.1.8. REMARK. In fact, we have proved the following stronger assertion (compare with [15], p. 253):

(i) *If $\hat{\varrho}$ is an arbitrary product metric in $C^n \times C^n$ that satisfies for every A_i, B_i, the condition*

$$\varrho_H(A_i, B_i) \leq \hat{\varrho}((A_1, A_2), (B_1, B_2)) \text{ for } i = 1, 2,$$

then Minkowski addition is uniformly continuous with respect to the metrics $\hat{\varrho}$ and ϱ_H.[3]

---

[1] That means a metric that induces the product topology.
[2] Product metrics are studied extensively in [34].
[3] See condition (2.2).

(ii) *Multiplication by an arbitrary nonnegative t,*

$$A \mapsto tA,$$

is uniformly continuous (with respect to ϱ_H).

2.2 Support hyperplane. The width

If E is a closed half-space in R^n and $H = \mathrm{bd}E$, then for any nonzero vector v orthogonal to H either $H + v \subset \mathrm{int}E$ or $(H + v) \cap E = \emptyset$. In the second case v is an *outer normal vector of E*.

2.2.1. DEFINITION. Let A be a closed, nonempty subset of R^n. A closed half-space E is a *support half-space of A* if

$$A \subset E \quad \text{and} \quad A \cap \mathrm{bd}E \neq \emptyset.$$

Then the hyperplane $H := \mathrm{bd}E$ is a *support hyperplane of A*, the set $A \cap H$ is a *support set*, every point of this set is a *support point*, and an outer normal vector v of the half-space E is an *outer normal vector of H*.

We shall prove the following.

2.2.2. THEOREM. *For every $A \in C^n$ and every $v \neq 0$ there is a unique support hyperplane of A with outer normal vector v.*

Proof. Let $\mathcal{H}$ be the set of hyperplanes orthogonal to v and let $L := \mathrm{lin}v$. For every $H_1, H_2 \in \mathcal{H}$,

$$\mathrm{dist}(H_1, H_2) = \|x_1 - x_2\|,$$

where $\{x_i\} = L \cap H_i$ for $i = 1, 2$ (compare (0.2)).

Obviously, $\mathrm{dist}|\mathcal{H} \times \mathcal{H}$ is a metric in $\mathcal{H}$, and the function $\hat{\pi} : \mathcal{H} \rightarrow L$ that assigns to any $H \in \mathcal{H}$ its orthogonal projection $\pi_L(H)$ is an isometry.

Since A is compact and nonempty, it follows that so is $\pi_L(A)$. Let

$$a \in \pi_L(A) \quad \text{and} \quad t_0 := \sup\{t \in R \mid a + t \cdot v \in \pi_L(A)\}.$$

Then evidently, $\hat{\pi}^{-1}(a + t_0 \cdot v)$ is the unique support hyperplane of A with outer normal vector v. $\qquad\square$

In view of Theorem 2.2.2, we may use the symbols $H(A, v)$ and $E(A, v)$, respectively, for the support hyperplane and the support half-space of a compact set A with v being an outer normal vector, and the symbol $A(v)$ for the support set $A \cap H(A, v)$ (Figure 2.1).

Applying 2.1.3, one can easily prove the following.

2.2.3. *For every $A_1, A_2 \in C^n$ and $v \neq 0$,*

$$H(A_1 + A_2, v) = H(A_1, v) + H(A_2, v)$$
$$(A_1 + A_2)(v) = A_1(v) + A_2(v).$$

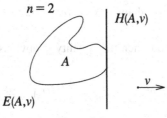

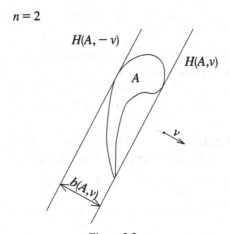

Figure 2.1.

2.2.4. DEFINITION. For $A \in \mathcal{C}^n$ and a nonzero vector v, let

$$b(A, v) := \text{dist}(H(A, v), H(A, -v)).$$

Then $b(A, v)$ is the *width of A in the direction of v* (Figure 2.2).

Figure 2.2.

Directly from 2.2.4 it follows that

2.2.5. *For every $A, B \in \mathcal{C}^n$,*

$$A \subset B \Longrightarrow \forall v \, b(A, v) \le b(B, v).$$

2.2.6. DEFINITION. For any $A \in \mathcal{C}^n$, let

$$d(A) := \inf_v b(A, v).$$

Then $d(A)$ is the *minimal width of A*.

Let us notice that the diameter of a set is its maximal width:

2.2.7. *For every $A \in \mathcal{C}^n$,*

$$\operatorname{diam} A = \sup_{v} b(A, v).$$

Proof. If A is a singleton, then the equality is obvious. Let $\operatorname{card} A \geq 2$. Since for every $x, y \in A$,

$$x \neq y \Longrightarrow \varrho(x, y) \leq \operatorname{dist}(H(A, y - x), H(A, x - y)) = b(A, x - y),$$

it follows that

$$\operatorname{diam} A \leq \sup_{v} b(A, v).$$

Let $v \neq 0$. If a and b are the support points for $H(A, v)$ and $H(A, -v)$, respectively, then

$$b(A, v) \leq \varrho(a, b) \leq \operatorname{diam} A.$$

Hence

$$\sup_{v} b(A, v) \leq \operatorname{diam} A. \qquad \square$$

2.2.8. DEFINITION. The mean value of the function $b \mid C^n \times S^{n-1}$ is referred to as the *mean width*: for every $A \in C^n$,

$$\bar{b}(A) := \frac{1}{\sigma(S^{n-1})} \int_{S^{n-1}} b(A, u) \, d\sigma(u).$$

We shall prove that the minimal width, diameter, and mean width are continuous; moreover, they are Lipschitz functions (see 2.2.10).

2.2.9. LEMMA. *For every* $A \in C^n$, $v \neq 0$, *and* $\delta > 0$,

$$b((A)_\delta, v) = b(A, v) + 2\delta.$$

Proof. Let x and y be support points for $H((A)_\delta, v)$ and $H((A)_\delta, -v)$. By 1.1.7, there exist $a, b \in A$ such that $x \in \{a\}_\delta$ and $y \in \{b\}_\delta$. Moreover, since $x, y \in \operatorname{bd}(A)_\delta$, it follows that

$$\|x - a\| = \delta = \|y - b\|.$$

It is easy to see that $H((A)_\delta, v)$ is also a support hyperplane (and thus a tangent hyperplane) of the ball $\{a\}_\delta$ at x, and a is a support point for $H(A, v)$. The situation is similar for the vector $-v$ and the point b. Thus the vectors $x - a$ and $y - b$ are parallel to v, and

$$b((A)_\delta, v) = \operatorname{dist}(H((A)_\delta, v), H((A)_\delta, -v))$$
$$= \operatorname{dist}(H(A, v), H(A, -v)) + 2\delta = b(A, v) + 2\delta. \qquad \square$$

2.2.10. THEOREM. *The functions* d, diam, $\bar{b} : C^n \to \mathrm{R}$ *are Lipschitz continuous.*

Proof. We may assume that $A \neq B$. Let $\delta = \varrho_H(A, B)$. Then $A \subset (B)_\delta$ and $B \subset (A)_\delta$, whence by 2.2.5 and 2.2.9,

$$b(A, v) \leq b(B, v) + 2\delta \quad \text{and} \quad b(B, v) \leq b(A, v) + 2\delta.$$

By 2.2.6–2.2.8, passing to the lower and the upper bound for $v \in S^{n-1}$, we obtain

$$|d(A) - d(B)| \leq 2\varrho_H(A, B) \quad \text{and} \quad |\mathrm{diam}A - \mathrm{diam}B| \leq 2\varrho_H(A, B);$$

Integrating over S^{n-1}, we obtain

$$|\bar{b}(A) - \bar{b}(B)| \leq 2\varrho_H(A, B).$$

$\square$

We close this section by proving simple theorems on two other real functions on C^n: the volume and the radius of the circumscribed ball with center 0.

2.2.11. THEOREM. *Let* $A, A_i \in C^n$, $i \in \mathbb{N}$, $A = \lim_H A_i$. *Then*

$$\limsup V_n(A_i) \leq V_n(A).$$

Proof. Since

$$\bigcap_{\varepsilon > 0} (A)_\varepsilon = A,$$

by the properties of any measure,

$$\lim_{\varepsilon \to 0} V_n((A)_\varepsilon) = V_n(A). \tag{2.3}$$

By the assumption and by the monotonicity of measure, we obtain

$$\forall \varepsilon > 0 \ \exists i_0 \ \forall i > i_0 \ V_n(A_i) \leq V_n((A)_\varepsilon);$$

Hence

$$\forall \varepsilon > 0 \ \limsup V_n(A_i) \leq V_n((A)_\varepsilon),$$

which, together with (2.3), completes the proof. $\square$

2.2.12. DEFINITION. For every $A \in C^n$, let

$$r_0(A) := \inf\{\alpha > 0 \mid A \subset \alpha \cdot B^n\}.$$

2.2.13. THEOREM. *The function* $r_0 : C^n \to \mathbb{R}$ *is a weak contraction.*
Proof. Let us notice that by 2.2.12, for every $X \in C^n$,

$$X \subset r_0(X)B^n.$$

Thus, as can be easily checked, for every $\varepsilon > 0$

$$(X)_\varepsilon \subset r_0(X)B^n + \varepsilon B^n = (r_0(X) + \varepsilon)B^n.$$

Hence

$$r_0((X)_\varepsilon) \leq r_0(X) + \varepsilon. \tag{2.4}$$

Let $\varepsilon := \varrho_H(A, B)$; then $A \subset (B)_\varepsilon$ and $B \subset (A)_\varepsilon$, whence by (2.4), $r_0(A) \leq r_0(B) + \varepsilon$ and $r_0(B) \leq r_0(A) + \varepsilon$. Therefore, $|r_0(A) - r_0(B)| \leq \varrho_H(A, B)$. $\square$

2.3 Convex sets

We are now concerned with convex subsets of R^n.

2.3.1. DEFINITION. A set $A \subset R^n$ is *convex* if for every pair of its points, $\{a, b\}$, the segment $\Delta(a, b)$ is contained in A.

The following statement is a direct consequence of 2.3.1.

2.3.2. *An affine image of a convex set is convex.*

The Minkowski operations preserve convexity:

2.3.3. (i) *If A_1 and A_2 are convex, then $A_1 + A_2$ is convex.*
(ii) *If A is convex, then for any $t \in R$ the set tA is convex.*

Proof. (i): Let $x, y \in A_1 + A_2$. Then $x = x_1 + x_2$ and $y = y_1 + y_2$ for some $x_i, y_i \in A_i$, $i = 1, 2$.
Since A_i is convex, it follows that $\Delta(x_i, y_i) \subset A_i$. Thus

$$
\begin{aligned}
\Delta(x, y) = & \quad \{(1 - t)x + ty \mid t \in [0, 1]\} \\
\subset & \quad \{(1 - t)x_1 + ty_1 \mid t \in [0, 1]\} + \{(1 - t)x_2 + ty_2 \mid t \in [0, 1]\} \\
\subset & \quad \Delta(x_1, y_1) + \Delta(x_2, y_2) \subset A_1 + A_2.
\end{aligned}
$$

(ii) follows directly from 2.3.2. ☐

Let us give a few simple examples:

2.3.4. EXAMPLES. (a) Every segment (with two endpoints, one endpoint, or without any endpoints) as well as any affine subspace of dimension $k \in \{0, \ldots, n\}$ is convex. In view of 2.3.2, it suffices to consider the segment $\Delta(0, e_1)$ on the axis x_1 and a subspace spanned by k axes of the coordinate system (for $k \neq 0$).

(b) Every ball is convex. Indeed, in view of 2.3.2, it suffices to show that B^n is convex. Let $a, b \in B^n$, i.e., $\|a\| \leq 1$ and $\|b\| \leq 1$. Then

$$
\forall t \in [0, 1] \ \|(1 - t)a + tb\| \leq (1 - t)\|a\| + t\|b\| \leq 1 - t + t = 1,
$$

whence $\Delta(a, b) \subset B^n$. ☐

By 2.1.5, 2.3.3, and 2.3.4(b) we infer that

2.3.5. *If A is convex, then for every $\varepsilon > 0$ its ε-hull $(A)_\varepsilon$ is convex.*

It is evident that the Minkowski operations restricted to convex sets share all the properties described in Section 2.1 for any subsets of R^n. Moreover, multiplication of a convex set by nonnegative scalars is distributive with respect to addition of scalars, i.e., in (2.1) the ball B^n can be replaced by an arbitrary convex set A:

2.3.6. PROPOSITION. *For every convex A and every $\alpha, \beta \geq 0$,*

$$
\alpha A + \beta A = (\alpha + \beta)A. \tag{2.5}
$$

Proof. If $\alpha = 0 = \beta$, then (2.5) has the form $\{0\} = \{0\}$. Let us assume that $\alpha + \beta > 0$.

The inclusion $\supset$ in (2.5) is true for arbitrary A.

$\subset$: Let $t := \frac{\beta}{\alpha+\beta}$. Then $t \in [0, 1]$ and $1 - t = \frac{\alpha}{\alpha+\beta}$; thus for every $a_1, a_2 \in A$,

$$\alpha a_1 + \beta a_2 = (\alpha + \beta)((1 - t)a_1 + ta_2) \in (\alpha + \beta)\Delta(a_1, a_2).$$

Hence $\alpha A + \beta A \subset (\alpha + \beta)A$, because $\Delta(a_1, a_2) \subset A$.   □

The following example proves that in 2.3.6 the assumption that A is convex is essential.

2.3.7. EXAMPLE. Let $A = S^{n-1}$. Then $0 \in A + A$, but $0 \notin 2A$, because $2A$ is the sphere with radius 2 and center 0. Thus $A + A \not\subset 2A$.   □

It is easy to check that

2.3.8. *The closure of any convex set is convex.*

We shall prove the following.

2.3.9. PROPOSITION. *For every closed subset X of R^n the following conditions are equivalent:*
(i) *X is convex,*
(ii) *$\forall a, b \in X$ $\frac{1}{2}(a + b) \in X$.*

Proof. The implication (i) $\Longrightarrow$ (ii) is evident.

(ii) $\Longrightarrow$ (i): Let $a, b \in X$. From (ii) it follows that the set $X \cap \Delta(a, b)$ is dense in $\Delta(a, b)$, whence

$$\Delta(a, b) \subset \text{cl}X = X,$$

which completes the proof.   □

2.3.10. DEFINITION. Let $X \subset R^n$. A function $f : X \to R$ is *convex* if its epigraph,

$$\{(x, t) \in X \times R : t \geq f(x)\},$$

is convex.

A function f is *concave* if $-f$ is convex.

Proof of the following statement is left to the reader (Exercise 2.5):

2.3.11. PROPOSITION. *Let X be a convex subset of R^n. For every continuous function $f : X \to R$ the following conditions are equivalent:*
(i) *f is convex,*
(ii) *$\forall x, y \in X$ $\forall t \in [0, 1]$ $f((1 - t)x + ty) \leq (1 - t)f(x) + tf(y)$,*
(iii) *$\forall x, y \in X$ $f(\frac{1}{2}(x + y)) \leq \frac{1}{2}(f(x) + f(y))$.*

2.4 Compact convex sets. Convex bodies

Let $\mathcal{K}^n$ be the class of nonempty compact convex subsets of R^n, and let $\mathcal{K}^n_0$ be the class of *convex bodies*, i.e., compact convex subsets of R^n with nonempty interior.

Since affine maps of $\mathbf{R}^n$ into itself are continuous and affine automorphisms (i.e., affine bijections) of $\mathbf{R}^n$ are homeomorphisms, from 2.3.2 we deduce the following.

2.4.1. COROLLARY. *The class $\mathcal{K}^n$ is invariant under affine maps and $\mathcal{K}_0^n$ is invariant under affine automorphisms of $\mathbf{R}^n$.*

We shall prove

2.4.2. PROPOSITION. (i) *$\mathcal{K}^n$ is closed under the Minkowski operations.*
(ii) *$\mathcal{K}_0^n$ is closed under Minkowski addition; moreover, if $A_1 \in \mathcal{K}_0^n$ and $A_2 \in \mathcal{K}^n$, then $A_1 + A_2 \in \mathcal{K}_0^n$.*
(iii) *multiplication by any nonzero scalar preserves $\mathcal{K}_0^n$.*

Proof. (i): By 2.1.7, the Minkowski operations preserve compactness, and by 2.3.3 they preserve convexity.
(ii): Let $A_1 \in \mathcal{K}_0^n$ and $A_2 \in \mathcal{K}^n$. By (i), it suffices to verify

$$\mathrm{int}(A_1 + A_2) \neq \emptyset. \tag{2.6}$$

Since $\mathrm{int}A_1 \neq \emptyset$, it follows that $\mathrm{int}(A_1 + x) \neq \emptyset$ for every x (because the translations are homeomorphisms). This together with the equality $A_1 + A_2 = \bigcup\{A_1 + x \mid x \in A_2\}$ yields (2.6).
(iii) also follows from (i) and topological invariance of interior (because the homotheties are homeomorphisms). $\qquad\qquad\square$

As a direct consequence of 2.3.5 and 2.4.2(ii),(iii), we obtain the following.

2.4.3. $A \in \mathcal{K}^n \implies (A)_\varepsilon \in \mathcal{K}_0^n$.

Let us note that $\mathcal{K}_0^n$ is not closed in $\mathcal{K}^n$ (so all the more in $\mathcal{C}^n$): e.g., a sequence of concentric balls B_k with radii $\frac{1}{k}$ is Hausdorff convergent to a singleton.
In the next chapter we shall prove that $\mathcal{K}^n$ is closed in $\mathcal{C}^n$ (Corollary 3.2.13).

2.5 Hyperplanes

This section is of auxiliary character.

Let $\mathcal{E}^n$ be the set of all the hyperplanes in $\mathbf{R}^n$. It can be parametrized by means of the function $\phi : S^{n-1} \times \mathbf{R}_+ \to \mathcal{E}^n$ defined by the formula

$$\phi(v, t) := \{x \in \mathbf{R}^n \mid x \circ v = t\}. \tag{2.7}$$

Evidently,

2.5.1. (a) *If $E = \phi(v, t)$, then*

$$t = \mathrm{dist}(0, E) \quad \text{and} \quad tv = \pi_E(0).$$

(Figure 2.3).

(b) *The function ϕ is a surjection, $\phi|(S^{n-1} \times (0, \infty))$ is an injection, and*

$$\forall v \in S^{n-1} \quad \phi(v, 0) = \phi(-v, 0).$$

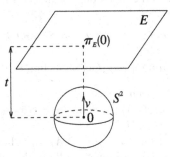

Figure 2.3.

We introduce a topology in $\mathcal{E}^n$, defining the limit (Exercise 2.6):

2.5.2. DEFINITION. For any sequence of hyperplanes $(E_k)_{k\in N}$, and for any $E \in \mathcal{E}^n$,

$$E = \lim_k E_k \Longleftrightarrow$$

$$\exists v, (v_k)_{k\in N}, t, (t_k)_{k\in N} \quad v = \lim_k v_k, \quad t = \lim_k t_k, \quad E = \phi(v, t), \quad E_k = \phi(v_k, t_k).$$

Directly from 2.5.2 it follows that (compare with Exercise 2.7)

2.5.3. *For every isometry $f : \mathbf{R}^m \to \mathbf{R}^m$ and arbitrary sequence $(E_k)_{k\in N}$ in $\mathcal{E}^m$,*

$$E = \lim E_k \Longrightarrow f(E) = \lim f(E_k).$$

Since every hyperplane in $\mathbf{R}^n$ is isometric to $\mathbf{R}^{n-1}$, the assertion 2.5.3 allows one to extend Definition 2.5.2: the set $\mathcal{E}^n$ can be replaced by the set of $(n-2)$-dimensional affine subspaces of some hyperplane.

2.5.4. PROPOSITION. *Let $E = \lim_k E_k$ for some $E, E_k \in \mathcal{E}^n$. Then*
(i) $(x_k)_{k\in N} \in \mathbf{P}_{k=1}^\infty E_k$, $x = \lim_k x_k \Longrightarrow x \in E$.
(ii) $\forall x \in E \; \exists (x_k)_{k\in N} \in \mathbf{P}_{k=1}^\infty E_k$, $x = \lim x_k$.

Proof. Let $E_k = \phi(v_k, t_k)$, $E = \phi(v, t)$, $v = \lim v_k$, $t = \lim t_k$.
(i): If $x_k \in E_k$, then by (2.7), $x_k \circ v_k = t_k$; thus $x \circ v = t$ by the continuity of scalar product. Hence $x \in E$.
(ii): Let $x \in E$. Then $x \circ v = t$.
In view of 2.5.3 we may assume that $x = 0 \in E$, whence $t = 0$. Let $x_k := \pi_{E_k}(0)$ for every k. Then $x_k = s_k v_k \in E_k$ for some $s_k \geq 0$; thus $t_k = x_k \circ v_k = s_k$, and hence $x_k = t_k v_k$ for every k. Passing to the limit, we obtain $\lim x_k = 0$, which completes the proof. $\qquad\square$

Proof of the next proposition is left to the reader (Exercise 2.8):

2.5.5. PROPOSITION. *If H, E, $E_k \in \mathcal{E}^n$ for $k \in \mathbf{N}$ and $H \cap E \neq H \neq H \cap E_k$ for every k, then*

$$E = \lim E_k \Longrightarrow E \cap H = \lim(E_k \cap H).$$

We shall now prove two theorems on a sequence of intersections of convex bodies by hyperplanes.

2.5.6. THEOREM. *Let $A \in \mathcal{K}^n$. If $E_k \in \mathcal{E}^n$ and $E_k \cap \operatorname{int} A \neq \emptyset$ for $k = 0, 1, \dots$, then*

$$E_0 = \lim E_k \Longrightarrow E_0 \cap A = \lim_{H}(E_k \cap A).$$

Proof. Let $E_0 = \lim E_k$. Then
(a) $\forall (x_{i_k})_{k \in \mathbf{N}} \in \mathbf{P}_{k=1}^{\infty}(E_{i_k} \cap A)$ $x_0 = \lim x_k \Longrightarrow x_0 \in E_0 \cap A$,
and
(b) $\forall x_0 \in E_0 \cap A \; \exists (x_k)_{k \in \mathbf{N}} \in \mathbf{P}_{k=1}^{\infty}(E_k \cap A)$ $x_0 = \lim x_k$.

Indeed, since A is closed, condition (a) follows directly from 2.5.4 (i); let us prove (b) (Figure 2.4). Obviously, we may assume that $E_0, E_1, \dots$ are pairwise distinct.

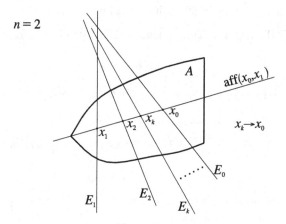

Figure 2.4.

Let $x_0 \in E_0 \cap A$, $x_1 \in E_1 \cap \operatorname{int} A \setminus \{x_0\}$, and let x_k be the intersection point of E_k and $\operatorname{aff}\{x_0, x_1\}$. Then $x_k \in A$.
Let $v_k \perp E_k$, $v_0 = \lim v_k$, and let $\beta_k = \angle((x_1 - x_0), v_k)$ for $k = 0, 1, \dots$.
It is easy to see that $\beta = \lim \beta_k$ and

$$\|x_k - x_0\| = \operatorname{dist}(x_0, E_k) \cdot \frac{1}{\sin \beta_k}.$$

Hence, in view of 2.5.1 and 2.5.2, $x_0 = \lim x_k$. This proves (b).

By Theorem 1.8.7 in [64] (compare Exercise 1.2), from (a) and (b) it follows that

$$E_0 \cap A = \lim_{H}(E_k \cap A).$$

This completes the proof. $\qquad\square$

2.5.7. THEOREM.[4] *Let* $A_k \in \mathcal{K}^n$ *for* $k \in \mathbb{N} \cup \{0\}$ *and* $E \in \mathcal{E}^n$. *If* $E \cap \mathrm{int} A_k \neq \emptyset$ *for* $k = 0, 1, \ldots,$ *then*

$$A_0 = \lim_H A_k \implies A_0 \cap E = \lim_H (A_k \cap E).$$

Proof. It suffices to prove that for any $Y \in \mathcal{K}^n$,

$$\forall \varepsilon > 0 \; \exists \delta > 0 \; (E \cap \mathrm{int} Y \neq \emptyset \implies (Y)_\delta \cap E \subset (Y \cap E)_\varepsilon). \qquad (2.8)$$

Indeed, if Y satisfies (2.8), then for every $X \in \mathcal{K}^n$,

$$X \subset (Y)_\delta \implies X \cap E \subset (Y)_\delta \cap E \implies X \cap E \subset (Y \cap E)_\varepsilon;$$

thus for $X := A_k$ and $Y := A_0$, since

$$\forall \delta > 0 \; \exists k_0 \; \forall k > k_0 \;\; A_k \subset (A_0)_\delta,$$

it follows that

$$\forall \varepsilon > 0 \; \exists k_0 \; \forall k > k_0 \;\; A_k \cap E \subset ((A_0)_\delta \cap E)_\varepsilon.$$

For $X := A_0$ and $Y := A_k$ the proof is analogous. Thus (2.8) yields the conclusion.

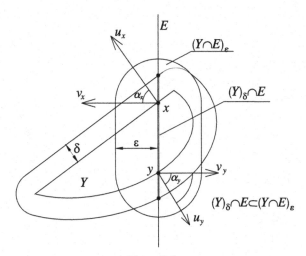

Figure 2.5.

Let $Y \in \mathcal{K}^n$. We prove (2.8). For every $x \in \mathrm{relbd}(Y \cap E)$ we choose an outer normal vector $u_x \in S^{n-1}$ for Y at the point x. Let $v_x \in S^{n-1}$ be the normal vector of the hyperplane E for which

---

[4] See [41], Theorem 14.3, which is a little stronger: the assumption that E has a nonempty intersection with the interior of A_k is made only for $k = 0$.

$$\alpha_x := \angle(v_x, u_x) \leq \frac{\pi}{2}.$$

Let us note that α_x depends not only on x, but also on the choice of u_x; however, since $E \cap \text{int} Y \neq \emptyset$, it follows that $\alpha_x \neq 0$ independently of the choice of u_x. Now let $\varepsilon > 0$. We define:

$$\delta := \inf_{x \in \text{relbd}(Y \cap E)} \inf_{u_x} \varepsilon \sin \alpha_x.$$

The set $\text{relbd}(Y \cap E)$ is compact; thus, if $\delta = 0$, then $\alpha_x = 0$ for some x. Hence $\delta > 0$. It can be seen that $Y_\delta \cap E \subset (Y \cap E)_\varepsilon$ (compare with Figure 2.5), which proves (2.8). $\qquad \square$

3
Basic Properties of Convex Sets

3.1 Convex combinations

In affine geometry the *affine combination of points* $a_1, \ldots, a_k \in \mathbf{R}^n$ with coefficients $t_1, \ldots, t_k \in \mathbf{R}$ satisfying $\sum_{i=1}^{k} t_i = 1$ is defined as the point $\sum_{i=1}^{k} t_i a_i$.

Further, the *affine subspace* aff$A$ *generated by a nonempty subset* A of $\mathbf{R}^n$ is understood as the set of affine combinations of points from A. It is also called the *affine hull of* A. We shall deal with counterparts of those notions in convex geometry: the convex combination and the convex hull.

3.1.1. DEFINITION. (i) Let $t_1, \ldots, t_k \in [0, 1]$ and $\sum_{i=1}^{k} t_i = 1$. The point

$$c(a_1, \ldots, a_k; t_1, \ldots, t_k) := \sum_{i=1}^{k} t_i a_i$$

is the *convex combination of points* $a_1, \ldots, a_k$ *with coefficients* $t_1, \ldots, t_k$.

(ii) For any $A \subset \mathbf{R}^n$

$$C(A) := \{c(a_1, \ldots, a_k; t_1, \ldots, t_k) \mid a_1, \ldots, a_k \in A,$$
$$t_1, \ldots, t_k \in [0, 1], \ k \in \mathbf{N}\}.$$

Thus, evidently, $C(A) \subset$ aff$A$.

If $\{a_1, \ldots, a_k\}$ is affine independent, then the set $C(\{a_1, \ldots, a_k\})$ is a *simplex*, and the points $a_1, \ldots, a_k$ are its *vertices*. A simplex with vertices $a_1, \ldots, a_k$ will be denoted by $\Delta(a_1, \ldots, a_k)$.

Hence (in accordance with the given notation), a simplex $\Delta(a_1, a_2)$ is the segment with endpoints a_1, a_2. A simplex $\Delta(a_1, a_2, a_3)$ is the triangle with vertices a_1, a_2, a_3.

It is easy to verify that

3.1.2. *For every $A \neq \emptyset$ the set $C(A)$ is convex.*

The following simple statement gives a parametric description of a convex set.

3.1.3. PROPOSITION. *For every nonempty subset A of R^n the following conditions are equivalent:*
(i) *A is convex,*
(ii) *$C(A) = A$.*

Proof. The implication (ii) $\Longrightarrow$ (i) follows directly from 3.1.2.
(i) $\Longrightarrow$ (ii): Let us note that for every A,

$$A \subset C(A). \tag{3.1}$$

Assume that A is convex. It suffices to show that for any $k \in N$,

$$a_1, \ldots, a_k \in A, \; t_1, \ldots, t_k \in [0, 1], \text{ and } \sum t_i = 1$$
$$\Longrightarrow c(a_1, \ldots, a_k; t_1, \ldots, t_k) \in A. \tag{3.2}$$

For $k = 1$ condition (3.2) is a tautology:

$$\forall a_1 \in A \; a_1 \in A.$$

Let $k \geq 2$. Assume that (3.2) is true for $k - 1$; let $a_1, \ldots, a_k \in A, \; t_1, \ldots, t_k \in [0, 1]$, and $\sum_{i=1}^{k} t_i = 1$.
If $t_k = 1$, then $c(a_1, \ldots, a_k; t_1, \ldots, t_k) = a_k \in A$. Let $t_k < 1$ and

$$t_i' = \frac{t_i}{1 - t_k} \text{ for } i = 1, \ldots, k - 1.$$

Then $\sum_{i=1}^{k-1} t_i' = 1$ and

$$c(a_1, \ldots, a_k; t_1, \ldots, t_k) = (1 - t_k)c(a_1, \ldots, a_{k-1}, t_1', \ldots, t_{k-1}') + t_k a_k \in A$$

by the inductive assumption. Thus (3.2) is true for k; this completes the proof. $\square$

The function C is Minkowski additive:

3.1.4. *For any nonempty A, B,*

$$C(A + B) = C(A) + C(B).$$

Proof.[1] $\subset$: Let $x \in C(A + B)$; thus x is a convex combination of some points in $A + B$, i.e., there exist $t_1, \ldots, t_k \in [0, 1], a_1, \ldots, a_k \in A$, and $b_1, \ldots, b_k \in B$ such that $\sum_i t_i = 1$ and

---

[1] Compare [64].

$$x = \sum_{i=1}^{k} t_i(a_i + b_i).$$

Hence $x \in C(A) + C(B)$.

$\supset$: Let now $x \in C(A) + C(B)$; then there exist $t_1, \ldots, t_k, s_1, \ldots, s_l \in [0, 1]$ such that $\sum t_i = 1 = \sum s_j$ and

$$x = \sum t_i a_i + \sum s_j b_j.$$

Of course, we may assume that $k = l$. Let $a := \sum t_i a_i$ and $b := \sum s_j b_j$. Then

$$x = \left(\sum s_j\right) a + \left(\sum t_i\right) b = \sum_{i,j} t_i s_j (a_i + b_j).$$

Since

$$\sum_{i,j} t_i s_j = \sum_i t_i \left(\sum_j s_j\right) = \left(\sum t_i\right)\left(\sum s_j\right) = 1,$$

it follows that $x \in C(A + B)$.   □

The following Carathéodory theorem says that the set $C(A)$ is generated by affine independent subsets of A; thus $C(A)$ is the union of simplices with vertices in A:

3.1.5. THEOREM. *Let $A \subset \mathbf{R}^n$. For every $x \in C(A)$ there exists an affine independent subset $\{a_1, \ldots, a_k\} \subset A$ such that $x \in \Delta(a_1, \ldots, a_k)$.*

Proof.[2] Let

$$k := \min\{m \in \mathbf{N} \mid x \in C(\{x_1, \ldots, x_m\}), \; x_i \in A, \; i = 1, \ldots, m\}.$$

Then there exist $a_1, \ldots, a_k \in A$ such that $x \in C(\{a_1, \ldots, a_k\})$, whence there exist $t_1, \ldots, t_k \in [0, 1]$ such that

$$\sum t_i = 1 \text{ and } x = \sum_{i=1}^{k} t_i a_i.$$

Suppose the set $\{a_1, \ldots, a_k\}$ is affine dependent, i.e., one of its points is an affine combination of the others. Equivalently, there exists $(s_1, \ldots, s_k) \neq (0, \ldots, 0)$, such that $\sum_{i=1}^{k} s_i a_i = 0$ and $\sum_{i=1}^{k} s_i = 0$. It is evident that at least one of the numbers $s_1, \ldots, s_k$ is positive; thus the set $\{\frac{t_i}{s_i} \mid i \in \{1, \ldots, k\}, \; s_i > 0\}$ is nonempty. Let $\frac{t_m}{s_m}$ be the smallest number of this set and let $\alpha_i := t_i - \frac{t_m}{s_m} \cdot s_i$ for $i = 1, \ldots, k$. Notice that $x = \sum_{i=1}^{k} \alpha_i a_i$, all the coefficients in this combination are nonnegative, their sum is equal to 1, and $\alpha_m = 0$. Thus x is a convex combination of $k - 1$ points, contrary to the assumption that k is minimal.   □

---

[2] Compare [64], p. 3.

Another important result, with various applications, is the following Helly theorem (for a proof see [64], Theorem 1.1.6).

3.1.6. THEOREM. *Let $\mathcal{X}$ be a finite family of convex subsets of R^n. If for every $A_1, \ldots, A_{n+1} \in \mathcal{X}$ the set $\bigcap_{i=1}^{n+1} A_i$ is nonempty, then*

$$\bigcap \mathcal{X} \neq \emptyset.$$

From Theorem 3.1.6 we derive its different version. It concerns a family of arbitrary cardinality, but the elements of this family are assumed to be compact.

3.1.7. THEOREM. *Let $\mathcal{X} \subset \mathcal{K}^n$. If for every $A_1, \ldots, A_{n+1} \in \mathcal{X}$ the set $\bigcap_{i=1}^{n+1} A_i$ is nonempty, then*

$$\bigcap \mathcal{X} \neq \emptyset.$$

Proof. In view of 3.1.6, every finite subfamily of $\mathcal{X}$ has a nonempty intersection; i.e., the family $\mathcal{X}$ has *the finite intersection property* (compare with [15], p. 123). Let $A_0 \in \mathcal{X}$ and let

$$\mathcal{X}_0 := \{A_0 \cap A \mid A \in \mathcal{X}\}.$$

Then $\mathcal{X}_0 \subset \mathcal{K}^n$ and $\bigcup \mathcal{X}_0 = A_0$, whence $\mathcal{X}_0$ consists of closed subsets of the compact space A_0. The family $\mathcal{X}_0$ has the finite intersection property as well, because for every $A_1, \ldots, A_k \in \mathcal{X}$,

$$\bigcap_{i=1}^{k}(A_0 \cap A_i) = \bigcap_{i=0}^{k} A_i \neq \emptyset.$$

Hence by Theorem 3.1.1 in [15], the family $\mathcal{X}_0$ has nonempty intersection, and thus $\bigcap \mathcal{X} \neq \emptyset$, because $\bigcap \mathcal{X}_0 \subset \bigcap \mathcal{X}$.   □

3.2 Convex hull

The following statement is a direct consequence of Definition 2.3.1.

3.2.1. PROPOSITION. *The intersection of an arbitrary family of convex sets is convex.*

3.2.2. DEFINITION. For any subset A of R^n, let $\mathcal{F}(A)$ be the family of all convex sets containing A. The intersection of $\mathcal{F}(A)$ is called the *convex hull of A*:

$$\operatorname{conv} A := \bigcap \mathcal{F}(A).$$

In view of 3.2.1, $\operatorname{conv} A$ is convex; it is the minimal (with respect to inclusion) convex subset of R^n containing A (see Figure 3.1).

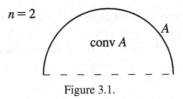

$n = 2$

A

conv A

Figure 3.1.

3.2.3. PROPOSITION. *For every A, $B \subset R^n$,*

$$A \subset B \Longrightarrow \operatorname{conv} A \subset \operatorname{conv} B.$$

Proof. It suffices to observe that

$$A \subset B \Longrightarrow \mathcal{F}(B) \subset \mathcal{F}(A). \qquad \square$$

We shall now prove that the convex hull coincides with the set of convex combinations:

3.2.4. THEOREM. *For every $A \neq \emptyset$*

$$\operatorname{conv} A = C(A).$$

Proof. By (3.1), A is a subset of $C(A)$. Since by 3.1.2, the set $C(A)$ is convex, it follows that

$$\operatorname{conv} A \subset C(A).$$

The operation C is increasing with respect to inclusion, whence $C(A) \subset C(\operatorname{conv} A)$. Thus, by the implication (i) $\Longrightarrow$ (ii) in 3.1.3,

$$C(A) \subset \operatorname{conv} A,$$

because conv A is convex. $\qquad \square$

From 3.2.4 combined with 3.1.3 we obtain the following characterization of convex sets.

3.2.5. COROLLARY. *A subset A of R^n is convex if and only if $A = \operatorname{conv} A$.*

The following statement is a direct consequence of 3.1.4 combined with 3.2.4.

3.2.6. COROLLARY. *For every A, $B \subset R^n$,*

$$\operatorname{conv}(A + B) = \operatorname{conv} A + \operatorname{conv} B.$$

From 3.2.6 we derive

3.2.7. PROPOSITION. conv $((A)_\varepsilon) = (\operatorname{conv} A)_\varepsilon$.
Proof. We apply in turn 2.1.5, 3.2.6, 2.3.4 (b), and 2.1.5:

$$\operatorname{conv}((A)_\varepsilon) = \operatorname{conv}(A + \varepsilon B^n) = \operatorname{conv} A + \operatorname{conv}(\varepsilon B^n)$$
$$= \operatorname{conv} A + \varepsilon B^n = (\operatorname{conv} A)_\varepsilon. \qquad \square$$

• Let us note that the convex hull of a closed set need not be closed. For example, let $A = \{(x_1, 0) \in \mathbb{R}^2 \mid 0 \leq x_1 \leq 1\} \cup \{(0, x_2) \in \mathbb{R}^2 \mid x_2 \geq 0\}$; then $\operatorname{conv} A = \{(x_1, x_2) \in \mathbb{R}^2 \mid 0 \leq x_1 < 1 \text{ and } x_2 \geq 0\} \cup \{(1, 0)\}$, whence $\operatorname{conv} A$ is not closed in $\mathbb{R}^2$, though A is closed.

However, the following holds.

3.2.8. THEOREM. $A \in \mathcal{C}^n \implies \operatorname{conv} A \in \mathcal{C}^n$.

Proof. Let, as above, $\mathcal{F}(A)$ be the family of convex sets in $\mathbb{R}^n$ containing A and let

$$\mathcal{F}_0(A) := \mathcal{F}(A) \cap \mathcal{C}^n.$$

On the one hand, by 3.2.1,

$$\operatorname{conv} A \subset \bigcap \mathcal{F}_0(A). \tag{3.3}$$

On the other hand, for every $X \in \mathcal{F}(A)$ there exists a set $X_0 \in \mathcal{F}_0(A)$ contained in X. For example, let $X_0 = B \cap \operatorname{cl} X$, where B is a ball containing A (then X_0 is compact, since it is a closed and bounded subset of $\mathbb{R}^n$, and is convex by 2.3.4, 2.3.8, and 3.2.1). Hence

$$\operatorname{conv} A \supset \bigcap \mathcal{F}_0(A). \tag{3.4}$$

By (3.3) and (3.4), the set $\operatorname{conv} A$ is the intersection of compact subsets of $\mathbb{R}^n$, whence it is compact. $\qquad\square$

In view of Theorem 3.2.8, the function conv may be considered as a function of $\mathcal{C}^n$ into itself, or as a function from $\mathcal{C}^n$ into $\mathcal{K}^n$. By 3.2.5, it maps $\mathcal{C}^n$ onto $\mathcal{K}^n$:

3.2.9. THEOREM. *The function* $\operatorname{conv} : \mathcal{C}^n \to \mathcal{K}^n$ *is surjective.*

We shall prove

3.2.10. THEOREM. *The function* $\operatorname{conv} : \mathcal{C}^n \to \mathcal{K}^n$ *is a weak contraction.*

Proof. Let $A_1, A_2 \in \mathcal{C}^n$. We have to prove that

$$\varrho_H(\operatorname{conv} A_1, \operatorname{conv} A_2) \leq \varrho_H(A_1, A_2). \tag{3.5}$$

Since condition (3.5) is symmetric with respect to A_1, A_2, it suffices to prove the inequality

$$\inf\{\delta > 0 \mid \operatorname{conv} A_1 \subset (\operatorname{conv} A_2)_\delta\} \leq \inf\{\delta > 0 \mid A_1 \subset (A_2)_\delta\}. \tag{3.6}$$

Let us note that in view of 3.2.3 and 3.2.7,

$$A_1 \subset (A_2)_\delta \implies \operatorname{conv} A_1 \subset (\operatorname{conv} A_2)_\delta,$$

whence the set on the left-hand side of (3.6) contains the set on the right-hand side. This implies (3.6). $\qquad\square$

As a direct consequence of 3.2.10, we obtain

3.2.11. COROLLARY. *The function* $\operatorname{conv} : \mathcal{C}^n \to \mathcal{K}^n$ *is continuous.*

In view of 3.2.5 and 3.2.11, the function conv is a continuous map of the space C^n onto its subset $\mathcal{K}^n$, and $\mathrm{conv}\, A = A$ for every $A \in \mathcal{K}^n$. We can formulate it as follows:

3.2.12. COROLLARY. *The function* conv *is a retraction of* C^n *on* $\mathcal{K}^n$.

Since a retract of any space is closed in this space, from 3.2.12 we deduce the following.

3.2.13. COROLLARY. *The family* $\mathcal{K}^n$ *is closed in* C^n.

Evidently, every closed subspace of a finitely compact metric space is finitely compact. Therefore, as a consequence of Corollary 3.2.13 combined with Theorem 1.2.8, one obtains the following theorem, which is well known as the Blaschke Selection Theorem (compare with[64], p.!50).

3.2.14. THEOREM. *The space* $\mathcal{K}^n$ *is finitely compact.*

3.3 Metric projection

In Section 3.2 we characterized convex sets in terms of the operation conv (Corollary 3.2.5).

The following theorem gives a characterization of closed convex sets in terms of distance between a point and a set.[3]

3.3.1. THEOREM. ([54]) *For every subset A of* $\mathbf{R}^n$ *the following conditions are equivalent:*

(i) *A is convex, nonempty, and closed in* $\mathbf{R}^n$,

(ii) *for every* $x \in \mathbf{R}^n$ *there exists a unique nearest point in A, i.e., a unique* $a \in A$ *with*

$$\varrho(x, a) = \varrho(x, A).$$

Proof. (i) $\Longrightarrow$ (ii):

Let $x \in \mathbf{R}^n$. If $x \in A$, then of course, x is a unique point of A nearest to x.

Let $x \notin A$. Since A is closed and nonempty, it follows that there is at least one $a \in A$ with $\varrho(x, a) = \varrho(x, A)$. Suppose there are two points a_1 and a_2 such that

$$\varrho(x, a_i) = \varrho(x, A) \text{ for } i = 1, 2,$$

and let $c = \frac{1}{2}(a_1 + a_2)$. Then c is the midpoint of the base $\Delta(a_1, a_2)$ of the isosceles triangle $\Delta(a_1, x, a_2)$, whence $\varrho(x, c) < \varrho(x, A)$. But this is impossible, because $c \in A$ by the convexity assumption.

(ii) $\Longrightarrow$ (i):

From (ii) it follows that $A \neq \emptyset$. It also follows that A is closed; indeed, otherwise there exists $x \in \mathrm{cl}A \setminus A$, and by (ii), there is an $a \in A$ such that $\varrho(x, a) = \varrho(x, A)$; hence $a = x$, since $\varrho(x, A) = 0$ by 1.1.4.

---

[3] See footnote on page 11 of [64].

Up to now we have used only a part of condition (ii): the existence of a nearest point in A for every point x. The uniqueness of the nearest point is needed to prove that A is convex (we roughly follow the argument in [64], Theorem 1.2.4).

Suppose that A is not convex. Then there exist $x_1, x_2 \in A$ such that $\Delta(x_1, x_2) \cap (R^2 \setminus A) \neq \emptyset$. For $i = 1, 2$, let a_i be the point in $\Delta(x_1, x_2) \cap \mathrm{bd}\, A$ most distant from x_i. Then $\Delta(a_1, a_2) \cap \mathrm{int}\, A = \emptyset$. Let B_0 be a ball with center $p_0 := \frac{a_1 + a_2}{2}$, disjoint from A:

$$B_0 = B(p_0, r_0).$$

The set of radii of the balls containing B_0 and disjoint from A is bounded; let r_1 be its upper bound and let

$$B_1 = B(p_1, r_1).$$

Of course, $B_1 \cap A$ is a singleton $\{p\}$, where p is the unique point of A nearest to p_1.

If $\mathrm{bd}\, B_0 \cap \mathrm{bd}\, B_1 \neq \emptyset$, then these two spheres have exactly one point in common, p, which does not belong to A, because $r_1 > r_0$. Since $p_0 \in \Delta(p_1, p)$, the sphere $\mathrm{bd}\, B_1$ has at least two points in common with $\mathrm{bd}\, A$, i.e. p_1 has more than one nearest point in A, contrary to the uniqueness part of (ii).

Thus

$$\mathrm{bd}\, B_1 \cap \mathrm{bd}\, B_2 = \emptyset,$$

whence B_1 is not a maximal ball containing B_0 and disjoint from $\mathrm{int}\, A$, because translating B_1 by $\varepsilon \cdot (p_0 - p_1)$ for a sufficiently small ε, we can obtain a ball disjoint from A with a bigger radius; thus radius r_1 is not maximal, contrary to the assumption. $\qquad\square$

In view of 3.3.1 (more precisely, in view of the implication (i) $\Longrightarrow$ (ii)), for every nonempty, closed, and convex subset A of R^n there exists a function assigning to a point $x \in R^n$ the point of A nearest to x.

3.3.2. DEFINITION. Let A be a nonempty, closed, and convex subset of R^n. The function $\xi_A : R^n \to A$ defined by the condition

$$\xi_A(x) = a \iff \varrho(x, a) = \varrho(x, A)$$

is the *metric projection onto A* (Figure 3.2).

We shall prove the following.

3.3.3. LEMMA. *Let A be a nonempty, convex, and closed subset of R^n; let $x \in R^n \setminus A$ and $a = \xi_A(x)$. If a hyperplane H_0 satisfies the conditions*

$$a \in H_0 \quad \text{and} \quad u = x - a \perp H_0,$$

then $H_0 = H(A, u)$.

Proof. Since $a \in A \cap H_0$, by definition of support hyperplane (Definition 2.2.1) it suffices to prove that the set A lies on the other side of H_0 from the point x. Suppose the opposite; then there exists a point $b \in A \setminus \{a\}$ such that

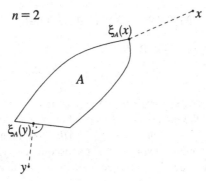

Figure 3.2.

$$0 < \angle(u, b - a) < \frac{\pi}{2}.$$

Since by the assumption, $\|x - b\| > \|x - a\|$, it follows that

$$\angle(a - b, x - b) < \frac{\pi}{2},$$

whence the height of the triangle $\Delta(x, a, b)$ passing through the vertex x intersects $\Delta(a, b)$ at some point c. Obviously, $c \in A$ and $\|x - c\| < \|x - a\|$, contrary to the assumption on a.   □

3.3.4. THEOREM. *For any nonempty, closed, convex subset A of $\mathbf{R}^n$, the metric projection ξ_A is a weak contraction.*

Proof. Let $x, y \in \mathbf{R}^n$, $a := \xi_A(x)$, and $b := \xi_A(y)$. We have to prove that

$$\|b - a\| \leq \|x - y\|. \tag{3.7}$$

Of course, (3.7) is satisfied if $a = b$. Assume that $a \neq b$. Let $v := a - b$ and let H_a and H_b be the hyperplanes orthogonal to v such that $a \in H_a$ and $b \in H_b$.

If $x \neq a$ and $y \neq b$, then from Lemma 3.3.3 it follows that a and b are support points for $H(A, x - a)$ and $H(A, y - b)$, respectively. Hence

$$v \circ (x - a) \geq 0 \quad \text{and} \quad v \circ (y - b) \leq 0. \tag{3.8}$$

It is obvious that (3.8) is also satisfied if $x = a$ or $y = b$.

By (3.8), the points x, y lie on different sides of the open strip bounded by the parallel hyperplanes H_a and H_b; this proves (3.7).   □

As a consequence of Theorem 3.3.4 and Lemma 3.3.3, we obtain the following.

3.3.5. THEOREM. *If A is a nonempty, closed, and convex subset of $\mathbf{R}^n$, then*

$$\xi_A(\mathbf{R}^n \setminus A) = \mathrm{bd}\,A.$$

Proof. The inclusion $\subset$ is obvious.

$\supset$: Let $a \in \text{bd}A$. Then there exists a sequence $(x_k)_{k \in \mathbb{N}}$ in $\mathbb{R}^n \setminus A$ convergent to a.

Let $a_k := \xi_A(x_k)$ for every k. By 3.3.4,

$$a = \lim a_k,$$

because $0 \leq \varrho(a, a_k) = \varrho(\xi_A(a), \xi_A(x_k)) \leq \varrho(a, x_k)$.

For some $\alpha > 0$ all the members of $(a_k)_{k \in \mathbb{N}}$ belong to $B(a, \alpha)$.

Let $L_k := a_k + \text{pos}(x_k - a_k)$ and $y_k \in L_k \cap \text{bd}B(a, \alpha)$; then evidently, $\xi_A(y_k) = a_k$ for every k. The sequence $(y_k)_{k \in \mathbb{N}}$ has a subsequence $(y_{i_k})_{k \in \mathbb{N}}$ convergent to a $y \in \text{bd}B(a, \alpha)$. Since the metric projections are continuous (compare 3.3.4), it follows that

$$\xi_A(y) = \lim \xi_A(y_{i_k}) = \lim a_{i_k} = a. \qquad \square$$

Let us note the following consequence of Theorem 3.3.5 combined with Lemma 3.3.3.

3.3.6. COROLLARY. *If A is a nonempty, convex, and closed subset of $\mathbb{R}^n$, then for every $a \in \text{bd}A$ there exists a support hyperplane of A passing through a.*

Moreover, the existence of a support hyperplane of A at every point of $\text{bd}A$ is equivalent to the assertion that A is nonempty, closed, and convex (compare Exercise 3.2). Thus we have another characterization of this class of sets.

Now we shall again make use of Lemma 3.3.3.

3.3.7. THEOREM. *For every nonempty closed subset A of $\mathbb{R}^n$, the following conditions are equivalent:*
(i) *A is convex,*
(ii) *A is the intersection of all its support half-spaces.*

Proof. The implication (ii)$\Longrightarrow$ (i) is evident.

(i) $\Longrightarrow$ (ii): Let $\mathcal{E}(A)$ be the family of support half-spaces of A. Obviously,

$$A \subset \bigcap \mathcal{E}(A).$$

Suppose $x \notin A$; then by 3.3.3, the hyperplane H_0 passing through the point $a = \xi_A(x)$ and perpendicular to $x - a$ is a support hyperplane for A. Thus x does not belong to the support half-space with boundary H_0, whence $x \notin \bigcap \mathcal{E}(A)$. $\qquad \square$

3.3.8. COROLLARY. *For any pair of disjoint sets $A_1, A_2 \in \mathcal{K}^n$, there exist disjoint closed half-spaces E_1, E_2 such that $A_i \subset E_i$ for $i = 1, 2$.*

Disjoint closed half-spaces E_1, E_2 whose existence is assured by 3.3.8 are said to *separate* the sets A_1, A_2.

3.4 Support function

The notion of support function is an important tool of convex geometry.

3.4.1. DEFINITION. Let $A \in \mathcal{K}^n$. The function $h_A : \mathbb{R}^n \to \mathbb{R}$ defined by the formula

$$h_A(x) := \sup\{a \circ x \mid a \in A\}$$

is the *support function* of A.

(We shall often write $h(A, \cdot)$ instead of $h_A(\cdot)$.)

The first part of the next theorem is a direct consequence of 3.4.1. It says that for every $A \in \mathcal{K}^n$ the support function of A is sublinear. For various proofs of the second part, about the existence and uniqueness of a compact convex A with $h_A = f$ for a given sublinear function $f : \mathbb{R}^n \to \mathbb{R}$, see [64], Theorem 1.7.1.

3.4.2. THEOREM. (i) *For every* $A \in \mathcal{K}^n$, *every* $x, y \in \mathbb{R}^n$, *and* $\alpha, \beta > 0$,

$$h_A(\alpha x + \beta y) \le \alpha h_A(x) + \beta h_A(y).$$

(ii) *For every sublinear function* $f : \mathbb{R}^n \to \mathbb{R}$ *there is a unique* $A \in \mathcal{K}^n$ *with* $h_A = f$.

Let us prove

3.4.3. PROPOSITION. *Let* $A \in \mathcal{K}^n$ *and* $u \in S^{n-1}$. *If* $a \in A \cap H(A, u)$, *then*

$$h_A(u) = a \circ u.$$

Proof. Let E_0 be the support half-space of A, with outer normal unit vector $u \in S^{n-1}$,

$$E_0 = E(A, u),$$

and let $a \in A \cap H(A, u)$. Obviously, for every $x \in E_0$,

$$(x - a) \circ u \le 0,$$

with equality for $x \in H(A, u)$. Since $A \subset E_0$, it follows that $x \circ u \le a \circ u$ for every $x \in A$ and

$$h_A(u) = \sup\{x \circ u \mid x \in A\} = a \circ u. \qquad \square$$

As we shall see, the support function restricted to S^{n-1} is positively linear with respect to the Minkowski operations:

3.4.4. THEOREM. *For any* $A_1, A_2 \in \mathcal{K}^n$, $t_1, t_2 \ge 0$, *and* $u \in S^{n-1}$,

$$h(t_1 A_1 + t_2 A_2, u) = t_1 h(A_1, u) + t_2 h(A_2, u).$$

Proof. Directly from 3.1.1, it follows that

$$h_{tA} = t \cdot h_A$$

for every $A \in \mathcal{K}^n$ and $t \ge 0$. Thus it suffices to prove that h is additive with respect to the first variable: for every $u \in S^{n-1}$,

$$h(A_1 + A_2, u) = h(A_1, u) + h(A_2, u). \tag{3.9}$$

To this end, let us note that for any support points a_1, a_2 of A_1, A_2, respectively, in the direction of the vector u, their sum $a_1 + a_2$ is a support point of the set $A_1 + A_2$ in the same direction (compare 2.2.3). Hence, (3.9) follows from 3.4.3. $\qquad\square$

The restriction of a support function to the unit sphere has the following geometric interpretation.

3.4.5. THEOREM. *For every $u \in S^{n-1}$,*

$$h_A(u) = \begin{cases} \operatorname{dist}(0, H(A, u)) & \text{if } 0 \in E(A, u) \\ -\operatorname{dist}(0, H(A, u)) & \text{if } 0 \notin E(A, u). \end{cases}$$

Proof. Let $a \in A \cap H(A, u)$. Then

$$H(A, u) = \{x \in \mathbf{R}^n \mid (x - a) \circ u = 0\},$$

whence $\operatorname{dist}(0, H(A, u)) = |a \circ u|$. Thus, by 3.4.3, the assertion follows. $\qquad\square$

3.4.6. COROLLARY. $h_A | S^{n-1} \leq h_B | S^{n-1} \iff A \subset B$.

Proof. From Theorem 3.4.5 it follows that for every $u \in S^{n-1}$,

$$h_A(u) \leq h_B(u) \iff E(A, u) \subset E(B, u).$$

It now suffices to apply 3.3.7. $\qquad\square$

We shall use 3.4.6 and 3.4.4 in the proof of the following cancellation law for Minkowski addition of compact convex subsets of $\mathbf{R}^n$:

3.4.7. THEOREM. *If $A_1, A_2, B \in \mathcal{K}^n$, then*

$$A_1 + B \subset A_2 + B \implies A_1 \subset A_2.$$

Proof. Let $A_1 + B \subset A_2 + B$. Then by 3.4.6 and 3.4.4, for every $u \in S^{n-1}$,

$$h(A_1, u) + h(B, u) \leq h(A_2, u) + h(B, u),$$

whence

$$h_{A_1} | S^{n-1} \leq h_{A_2} | S^{n-1}.$$

Thus, $A_1 \subset A_2$ by Corollary 3.4.6. $\qquad\square$

The following example shows that the convexity assumption in 3.4.7 is essential:

3.4.8. EXAMPLE. Let $A_1 = B = B^n$ and $A_2 = S^{n-1}$. Then $A_1 + B = A_2 + B = 2B$, while $A_1 \not\subset A_2$.

From 2.2.4 combined with 3.4.5 we deduce the following statement concerning width:

3.4.9. PROPOSITION. *For every $A \in K^n$ and $u \in S^{n-1}$*

$$b(A, u) = h(A, u) + h(A, -u).$$

Finally, let us note a useful relationship between the support function and the Hausdorff metric (compare with [64] Theorem 1.8.11).

3.4.10. THEOREM. *For every $A_1, A_2 \in K^n$,*

$$\varrho_H(A_1, A_2) = \sup_{u \in S^{n-1}} |h(A_1, u) - h(A_2, u)|.$$

Proof. By (1.4),

$$\varrho_H(A_1, A_2) = \inf\{\alpha > 0 \mid A_1 \subset A_2 + \alpha B^n \text{ and } A_2 \subset A_1 + \alpha B^n\}.$$

Hence, by 3.4.4 combined with 3.4.6,

$\varrho_H(A_1, A_2)$

$\quad = \inf\{\alpha > 0 \mid h_{A_1} | S^{n-1} \leq h_{A_2} | S^{n-1} + \alpha \text{ and } h_{A_2} | S^{n-1} \leq h_{A_1} | S^{n-1} + \alpha\}$

$\quad = \inf\{\alpha > 0 \mid \sup_{u \in S^{n-1}} |h(A_1, u) - h(A_2, u)| \leq \alpha\}$

$\quad = \sup_{u \in S^{n-1}} |h(A_1, u) - h(A_2, u)|.$ $\qquad\qquad\qquad\qquad\qquad \square$

4

Transformations of the Space $\mathcal{K}^n$ of Compact Convex Sets

Any continuous function $f : \mathrm{R}^n \rightarrow \mathrm{R}^n$ induces the function $f^* : C^n \rightarrow C^n$ that assigns to an $X \in C^n$ its image $f(X) \in C^n$:

$$f^*(X) := f(X).$$

There is an extensive literature concerning relationships between the properties of a function f and those of the induced function f^* (compare [37]).

4.1 Isometries and similarities

4.1.1. THEOREM. *For every $f : \mathrm{R}^n \rightarrow \mathrm{R}^n$ and $\lambda > 0$ the following conditions are equivalent*:
  (i) *f is a similarity with ratio λ of the space R^n*;
  (ii) *f^* is a similarity with ratio λ of the space C^n (with the Hausdorff metric)*.
  Proof. (i) $\Longrightarrow$ (ii):
Let f be a similarity with ratio λ. Obviously, for every $x \in \mathrm{R}^n$ and $Y \in C^n$,

$$\varrho(f(x), f(Y)) = \lambda \varrho(x, Y).$$

Thus, by Theorem 1.2.2,

$$\varrho_H(f(A), f(B)) = \max\left\{ \sup_{a \in A} \varrho(f(a), f(B)), \sup_{b \in B} \varrho(f(b), f(A)) \right\} = \lambda \varrho_H(A, B).$$

It remains to show that f^* is surjective. Let $Y \in C^n$ and $X := f^{-1}(Y)$. Then of course, $f^*(X) = Y$ and $X \in C^n$ (because f^{-1} is a similarity).

(ii) $\Longrightarrow$ (i): Assume f^* to be a similarity with ratio $\lambda > 0$; i.e., it is surjective and for every $A, B \in \mathcal{C}^n$,

$$\varrho_H(f^*(A), f^*(B)) = \lambda \varrho_H(A, B).$$

Hence f^* is injective, and for $A = \{a\}$ and $B = \{b\}$,

$$\|f(a) - f(b)\| = \varrho_H(f(A), f(B)) = \lambda \varrho_H(A, B) = \lambda \|a - b\|.$$

It remains to show that f is surjective. Let $y \in \mathbf{R}^n$; since $\{y\} = f^*(\{x\})$ for some $x \in \mathbf{R}^n$ and $f^*(\{x\}) = \{f(x)\}$; it follows that $y = f(x)$. $\square$

There exist isometric embeddings of the space $\mathcal{K}^n$ into $\mathcal{K}^n$ that are not induced by any transformations of $\mathbf{R}^n$:

4.1.2. EXAMPLE. Let $A \in \mathcal{K}^n$. The map

$$X \mapsto X + A$$

is an analogue of a translation of $\mathbf{R}^n$; we shall call it *translation (of $\mathcal{K}^n$ by A)*. From (1.4), 2.1.3, and 3.4.7 it follows that the restriction of this map to $\mathcal{K}^n$ is an isometric embedding of $\mathcal{K}^n$ into $\mathcal{K}^n$. Indeed, for every δ,

$$X + A \subset (Y + A) + \delta B^n \iff X \subset Y + \delta B^n,$$

whence $\varrho_H(X + A, Y + A) = \varrho_H(X, Y)$.

However, if A is not a singleton, then the translation by A is not surjective, and thus is not an isometry of $\mathcal{K}^n$. Moreover, it is not induced by any transformation of $\mathbf{R}^n$, since it does not preserve the family of singletons.

The following two interesting theorems together with Theorem 4.1.1 give a complete characterization of the isometries and the isometric embeddings of $\mathcal{C}^n$ and $\mathcal{K}^n$.

4.1.3. THEOREM. ([25]). *Every isometry of $\mathcal{C}^n$ onto $\mathcal{C}^n$ is induced by an isometry of $\mathbf{R}^n$.*[1]

4.1.4. THEOREM. ([26]). *Every isometric embedding F of $\mathcal{K}^n$ into $\mathcal{K}^n$ satisfies the condition*

$$F(X) = f^*(X) + A$$

for some isometry $f : \mathbf{R}^n \to \mathbf{R}^n$ and some $A \in \mathcal{K}^n$.

In the next section we shall consider a useful example of a continuous map of the space $\mathcal{K}_0^n$ into $\mathcal{K}_0^n$ that is not induced by any transformation of $\mathbf{R}^n$ (Theorems 4.2.14 and 4.2.16). We shall need the following statement, which is part of a theorem on characterization of induced maps due to the Charatoniks ([13]).[2]

---

[1] For convex sets this theorem was proved by R. Schneider in [61].

[2] Their theorem was proved for a class of sets that is slightly different from $\mathcal{C}^n$.

4.1.5. THEOREM. *Let $\mathcal{X}$ be a family of nonempty compact subsets of R^n that contains the set of singletons. If $F = f^* : \mathcal{X} \to \mathcal{X}$ for some $f : \mathrm{R}^n \to \mathrm{R}^n$ and $F' : \mathcal{X} \to \mathcal{X}$ preserves the set of singletons, preserves inclusion, and $F'(A) \subset F(A)$ for every $A \in \mathcal{X}$, then $F' = F$.*

Proof. Let $A \in \mathcal{X}$. For every $x \in A$ the set $F'(\{x\})$ is a singleton: $F'(\{x\}) = \{y\}$ for some $y \in F'(A)$. Since $F'(\{x\}) \subset F(\{x\}) = \{f(x)\}$, it follows that $F'(\{x\}) = \{f(x)\}$.

By the assumption, $F'(A) \subset f(A)$. It remains to prove that

$$f(A) \subset F'(A). \tag{4.1}$$

Let $y \in f(A)$; then $y = f(x)$ for some $x \in A$, whence $F'(A) \supset F'(\{x\}) = \{f(x)\}$. This proves (4.1). $\square$

4.2 Symmetrizations of convex sets. The Steiner symmetrization

4.2.1. DEFINITION. A function $F : \mathcal{K}^n \to \mathcal{K}^n$ is a *symmetrization* if there exists an affine subspace E such that for every $A \in \mathcal{K}^n$ the set $F(A)$ is symmetric with respect to E, that is, $\sigma_E(F(A)) = F(A)$.

This definition can be extended to all nonempty subsets of R^n.

We shall give several examples of symmetrizations. The most commonly used is the following one (compare [30]).

4.2.2. DEFINITION. Let H be a hyperplane in R^n. For every $x \in H$, let $L_x := H^\perp(x)$. For any $A \in \mathcal{K}^n$ and $x \in \pi_H(A)$, let a_x be the symmetry center of $A \cap L_x$, and define $v_x := x - a_x$ and $A_x := A \cap L_x + v_x$.

The set $S_H(A)$ is defined by the formula

$$S_H(A) := \bigcup_{x \in \pi_H(A)} A_x \tag{4.2}$$

(Figure 4.1). The function $A \mapsto S_H(A)$ is called the *Steiner symmetrization (with respect to H)*.

(Obviously, the set A_x is either a segment or a singleton.)

It is evident that the values of the function S_H are sets symmetric with respect to H. As we shall prove, these sets are compact and convex (Theorem 4.2.7).

It is easy to see that S_H preserves inclusion:

4.2.3. PROPOSITION. $A \subset B \implies S_H(A) \subset S_H(B)$.

The fixed points of S_H are the sets symmetric with respect to H:

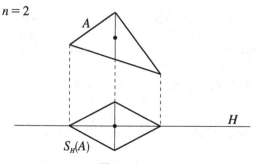

Figure 4.1.

4.2.4. PROPOSITION. $S_H(A) = A \Longleftrightarrow \sigma_H(A) = A$.

Proof. $\Longleftarrow$:

Let $\pi := \pi_H$. If H is a hyperplane of symmetry of A, then $A_x = A \cap L_x$ for every $x \in \pi(A)$, whence

$$S_H(A) = \bigcup_{x \in \pi(A)} A \cap L_x = A \cap \pi^{-1}\pi(A) = A,$$

because $A \subset \pi^{-1}\pi(A)$.

$\Longrightarrow$:

Let $S_H(A) = A$. Then by (4.2),

$$\bigcup_{x \in \pi(A)} (A \cap L_x + v_x) = \bigcup_{x \in \pi(A)} A \cap L_x;$$

thus $A \cap L_x + v_x = A \cap L_x$ for every $x \in \pi(A)$. Hence $v_x = 0$, i.e., $a_x \in H$ for every $x \in \pi(A)$. Therefore, H is a hyperplane of symmetry of A. $\square$

Directly from Definition 4.2.2 it follows that if two hyperplanes are parallel, then the images of a given set under the corresponding symmetrizations are translates of each other:

4.2.5. PROPOSITION. *If $v \perp H$ and $H' = H + v$, then for every $A \in \mathcal{K}^n$,*

$$S_{H'}(A) = S_H(A) + v.$$

Let us note the following.

4.2.6. PROPOSITION. (i) *If $v \perp H$, then for every $A \in \mathcal{K}^n$,*

$$S_H(A + v) = S_H(A).$$

(ii) *If $v \| H$, then for every $A \in \mathcal{K}^n$,*

$$S_H(A + v) = S_H(A) + v.$$

We shall prove

4.2.7. THEOREM.

(i) $A \in \mathcal{K}^n \implies S_H(A) \in \mathcal{K}^n$;

(ii) $A \in \mathcal{K}_0^n \implies S_H(A) \in \mathcal{K}_0^n$.

Proof. (i): Let $A \in \mathcal{K}^n$.

(a) The set A is bounded, whence $A \subset B(a, \alpha)$ for some $a \in H$ and $\alpha > 0$. Thus by 4.2.3 and 4.2.4,

$$S_H(A) \subset S_H(B(a, \alpha)) = B(a, \alpha);$$

hence $S_H(A)$ is bounded too.

(b) Since A is closed in $\mathbf{R}^n$, also $S_H(A)$ is closed. Indeed, let $y_i \in S_H(A)$ for $i \in \mathbf{N}$, $y = \lim y_i$, $x_i = \pi_H(y_i)$, $x = \pi_H(y)$, and let

$$z_i := y_i - v_{x_i} \text{ for } i \in \mathbf{N}, \text{ and } z := y - v_x.$$

Obviously, $z_i \in A$. It is easy to check that $\lim z_i = z$ (Exercise 4.2); thus $z \in A$ because A is closed. Hence $y \in S_H(A)$.

(c) Since A is convex, it follows that so is $S_H(A)$:

Take two points $y_1, y_2 \in S_H(A)$ and let $x_i := \pi_H(y_i)$ for $i = 1, 2$. Then $y_i \in A_{x_i}$ for $i = 1, 2$ (compare 4.2.2). Let

$$X = \text{conv}(A \cap L_{x_1} \cup A \cap L_{x_2}), \quad Y = \text{conv}(A_{x_1} \cup A_{x_2}).$$

Generally, the sets X and Y are trapezoids with their bases perpendicular to H. It is easy to verify that $Y = S_H(X)$. Since A is convex, it follows that $X \subset A$ and thus $Y \subset S_H(A)$ in view of 4.2.3. Therefore, $\Delta(y_1, y_2) \subset S_H(A)$.

(ii): By (i), it suffices to prove

$$\text{int} A \neq \emptyset \implies \text{int} S_H(A) \neq \emptyset.$$

If B is a ball contained in A, then by 4.2.5 and 4.2.3, the set $S_H(B)$ is a ball contained in $S_H(A)$. □

Volume is one of the important invariants of the Steiner symmetrization.

4.2.8. THEOREM. *For every* $A \in \mathcal{K}^n$,

$$V_n(S_H(A)) = V_n(A).$$

Proof. Applying Fubini's theorem twice, we obtain

$$V_n(S_H(A)) = \int_{\pi(A)} V_1(A_x) dV_{n-1}(x)$$

$$= \int_{\pi(A)} V_1(A \cap L_x) dV_{n-1}(x) = V_n(A).$$ □

The next two statements describe relationships between the Steiner symmetrization and the Minkowski operations.

4.2.9. THEOREM. *If* $0 \in H$, *then for every* $\lambda > 0$,

$$\lambda S_H(A) = S_H(\lambda A).$$

Proof. Since $0 \in H$, it follows that $\lambda H = H$ and $\lambda L_x = L_{\lambda x}$ for $x \in H$. Hence

$$\lambda(L_x \cap A) = L_{\lambda x} \cap (\lambda A),$$

because a homothety with nonzero ratio is a bijection. A homothety maps the midpoint of a segment to the midpoint of its image and preserves parallelism and congruence; hence

$$\lambda A_x = (\lambda A)_{\lambda x}$$

(Figure 4.2). By Thales Theorem, $\pi_H(\lambda A) = \lambda \pi_H(A)$; thus

$$\lambda S_H(A) = \bigcup_{x \in \pi_H(A)} \lambda A_x = \bigcup_{\lambda x \in \pi_H(\lambda A)} (\lambda A)_{\lambda x} = S_H(\lambda A). \qquad \square$$

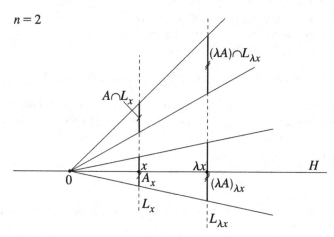

Figure 4.2.

4.2.10. THEOREM. *If* $0 \in H$, *then*

$$S_H(A + B) \supset S_H(A) + S_H(B).$$

Proof. Let $c \in S_H(A) + S_H(B)$; then $c = a + b$ for some $a \in S_H(A)$ and $b \in S_H(B)$. We have to prove that

$$c \in S_H(A + B). \tag{4.3}$$

We project a, b, c on the line $L_0 := H^\perp$; let

$$a' := \pi_{L_0}(a), \quad b' := \pi_{L_0}(b), \quad c' := \pi_{L_0}(c).$$

The map π_{L_0} is linear, whence $c' = a' + b'$. Consider the vectors $u := a' - a$, $v := b' - b$, and $w := c' - c$; let $A' := A + u$ and $B' := B + v$. Of course, $w = u + v$, whence $A' + B' = (A + B) + w$. We shall first show that

$$S_H(A' + B') \cap L_0 \supset S_H(A') \cap L_0 + S_H(B') \cap L_0. \qquad (4.4)$$

Let us note that $A' \cap L_0$ and $B' \cap L_0$ are segments (possibly degenerate) on the line L_0; further, $A' \cap L_0 + B' \cap L_0 \subset (A' + B') \cap L_0$, because L_0 is a linear subspace. Hence

$$V_1(A' \cap L_0) + V_1(B' \cap L_0) = V_1(A' \cap L_0 + B' \cap L_0) \leq V_1((A' + B') \cap L_0).$$

Consequently, by Definition 4.2.2, we obtain

$$V_1(S_H(A') \cap L_0) + V_1(S_H(B') \cap L_0) \leq V_1(S_H(A' + B') \cap L_0). \qquad (4.5)$$

Observe that all three segments mentioned in (4.5) are contained in L_0 and have the same midpoint; thus (4.5) implies (4.4).

In view of 4.2.6.(ii),

$$(S_H(A) + u) + (S_H(B) + v) = S_H(A') + S_H(B')$$

and

$$S_H(A + B) + w = S_H(A' + B'),$$

whence by (4.4),

$$(S_H(A) + u) \cap L_0 + (S_H(B) + v) \cap L_0 \subset S_H(A + B + w) \cap L_0$$

$$\subset S_H(A + B + w) = S_H(A + B) + w.$$

Thus $c + w \in S_H(A + B) + w$, which proves (4.3). $\qquad\qquad$ □

The following corollary is a direct consequence of 4.2.10 combined with 4.2.4.

4.2.11. COROLLARY. *For every $\varepsilon > 0$ and $A \in \mathcal{K}^n$,*

$$(S_H(A))_\varepsilon \subset S_H((A)_\varepsilon).$$

Let us notice that in Corollary 4.2.11 (and so in Theorem 4.2.10 as well) inclusion cannot be replaced by equality:

4.2.12. EXAMPLE. Let $n = 2$ and $H = \{(x_1, x_2) \mid x_2 = 0\}$; let $A = \Delta((0, 0), (2, 2))$ and $\varepsilon \leq \frac{1}{2}$. For $x = (1, 0)$, the set $L_x \cap S_H((A)_\varepsilon)$ is a segment with length $2\varepsilon\sqrt{2}$, while $L_x \cap (S_H(A))_\varepsilon$ is a segment with length 2ε. Hence $S_H((A)_\varepsilon) \neq (S_H(A))_\varepsilon$ (Figure 4.3 a,b).

It is easy to generalize this example to arbitrary n (Exercise 4.9).

As we already mentioned in Section 4.1, the function S_H is continuous on $\mathcal{K}_0^n$. For the proof we shall need the following simple lemma.

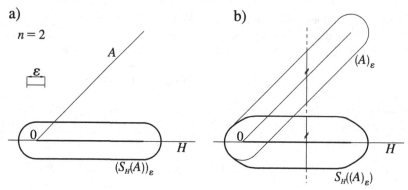

Figure 4.3.

4.2.13. LEMMA. *Let $A \in \mathcal{K}_0^n$, $\alpha, \beta, \delta > 0$, and $\lambda > 1$. If*

$$\alpha B^n \subset A \subset \beta B^n,$$

then

  (i) $A + \delta B^n \subset (1 + \frac{\delta}{\alpha})A$,

  (ii) $\lambda A \subset A + (\lambda - 1)\beta B^n$.

Proof. (i): $\alpha B^n \subset A \Longrightarrow \alpha\delta B^n \subset \delta A \Longrightarrow \delta B^n \subset \frac{\delta}{\alpha}A$. Thus in view of 2.3.6, we obtain (i).

  (ii): $A \subset \beta B^n \Longrightarrow \lambda A \subset A + (\lambda - 1)A \subset A + (\lambda - 1)\beta B^n$. $\square$

4.2.14. THEOREM. *For any hyperplane H in $\mathbf{R}^n$, the function $S_H \mid \mathcal{K}_0^n$ is continuous.*

Proof. Let $A, A_k \in \mathcal{K}_0^n$ for $k \in \mathbf{N}$ and let $A = \lim A_k$.[3] We have to show that

$$S_H(A) = \lim S_H(A_k). \tag{4.6}$$

Since translations are continuous with respect to the Hausdorff metric, and by 4.2.5, $S_{H+v}(X) = S_H(X) + v$ for every $X \in \mathcal{K}^n$ and $v \perp H$, without loss of generality we may assume that $0 \in H$.

Obviously, $0 \in \text{int}(A + u)$ for some unit vector u. Let

$$u = u_1 + u_2, \quad u_1 \| H, \ u_2 \perp H.$$

By 4.2.6,

$$S_H(A + u) = S_H(A + u_1) = S_H(A) + u_1$$

and analogously, for every k,

$$S_H(A_k + u) = S_H(A_k) + u_1;$$

thus we may also assume that $0 \in \text{int} A$.

---

[3] We write here lim instead of $\lim_H$.

Since $A = \lim A_k$, it follows that there is a function $\phi : (0, \infty) \to \mathbb{N}$ such that for every $\delta > 0$,

$$\forall k > \phi(\delta) \quad A \subset A_k + \delta B^n \quad \text{and} \quad A_k \subset A + \delta B^n. \tag{4.7}$_\delta$

The set A is bounded and $0 \in \mathrm{int} A$, whence for some $\alpha_1, \beta_1 > 0$,

$$\alpha_1 B^n \subset A \subset \beta_1 B^n. \tag{4.8}$$

Let

$$k_1 := \phi\left(\frac{\alpha_1}{2}\right), \quad \alpha := \frac{\alpha_1}{2}, \quad \beta := \frac{\alpha_1}{2} + \beta_1.$$

Then by (4.8),

$$\alpha B^n \subset A \subset \beta B^n, \tag{4.9}$$

and by (4.8) combined with (4.7)$_\alpha$, for $k > k_1$,

$$\alpha_1 B^n \subset A_k + \frac{\alpha_1}{2} B^n \subset A + \alpha_1 B^n \subset (\alpha_1 + \beta_1) B^n;$$

hence in view of the cancellation law 3.4.7,

$$\forall k > k_1 \quad \alpha B^n \subset A_k \subset \beta B^n. \tag{4.10}$$

Now let $\varepsilon > 0$ and $k_2 = \phi(\varepsilon \cdot \frac{\alpha}{\beta})$.
By condition (4.7)$_{\varepsilon \cdot \frac{\alpha}{\beta}}$,

$$A \subset A_k + \left(\varepsilon \cdot \frac{\alpha}{\beta}\right) B^n \quad \text{and} \quad A_k \subset A + \left(\varepsilon \cdot \frac{\alpha}{\beta}\right) B^n$$

for $k > k_2$. Hence by (4.9) and (4.10), from Lemma 4.2.13.(i) for $\delta = \varepsilon \cdot \frac{\alpha}{\beta}$, it follows that if $k > \max\{k_1, k_2\}$, then

$$A \subset \left(1 + \frac{\varepsilon}{\beta}\right) A_k \quad \text{and} \quad A_k \subset \left(1 + \frac{\varepsilon}{\beta}\right) A.$$

Thus by 4.2.3 (i) combined with 4.2.9,

$$S_H(A) \subset \left(1 + \frac{\varepsilon}{\beta}\right) S_H(A_k) \quad \text{and} \quad S_H(A_k) \subset \left(1 + \frac{\varepsilon}{\beta}\right) S_H(A).$$

This together with Lemma 4.2.13.(ii) for $\lambda = 1 + \frac{\varepsilon}{\beta}$ yields

$$S_H(A) \subset S_H(A_k) + \varepsilon B^n \quad \text{and} \quad S_H(A_k) \subset S_H(A) + \varepsilon B^n$$

for sufficiently large k. Thus the proof of condition (4.6) is complete. $\qquad\square$

Theorem 4.2.14 concerns the family of convex bodies $\mathcal{K}_0^n$. The following example shows that it cannot be generalized on $\mathcal{K}^n$, that is, the function $S_H : \mathcal{K}^n \to \mathcal{K}^n$ is not continuous.

4.2.15. EXAMPLE. Let $H = \{(0, t) \mid t \in R\} \subset R^2$. Consider the sequence of segments I_k in R^2 convergent to the segment I:

$$I_k = \Delta\left((0, 1), \left(\frac{1}{k}, 0\right)\right), \quad I = \Delta((0, 1), (0, 0)).$$

Obviously,

$$S_H(I_k) = \Delta\left((0, 0), \left(\frac{1}{k}, 0\right)\right), \quad S_H(I) = \Delta\left(\left(0, -\frac{1}{2}\right), \left(0, \frac{1}{2}\right)\right),$$

whence $\lim S_H(I_k) \neq S_H(I)$.

It is easy to modify this example for arbitrary n (Exercise 4.10).

4.2.16. THEOREM. *The Steiner symmetrization $S_H : \mathcal{K}^n \to \mathcal{K}^n$ is not induced by any transformation of R^n.*

Proof. According to 4.1.5, it suffices to find a function $F' : \mathcal{K}^n \to \mathcal{K}^n$ that preserves the family of singletons, preserves inclusion, and satisfies the following condition:

$$\forall A \in \mathcal{K}^n \ F'(A) \subset S_H(A) \quad \text{and} \quad \exists A \in \mathcal{K}^n \ F'(A) \neq S_H(A).$$

The simplest example of such a function is $(\pi_H)^*$. (Another example is $f^* S_H$, where f is an affine map such that H is the set of its fixed points and for every $x \in H$ the function $f|L_x$ is the homothety of L_x with ratio $\lambda < 1$.) $\square$

4.3 Other symmetrizations

The Steiner symmetrization is a particular case of symmetrization with respect to an arbitrary affine subspace (compare [20], p. 58).

4.3.1. DEFINITION. Let E be a k-dimensional affine subspace of R^n for some $k \in \{0, \ldots, n - 1\}$.

For any $A \in \mathcal{K}^n$ and $x \in \pi_E(A)$ let A_x be the ball in $E^\perp(x)$ with dimension equal to $\dim(A \cap E^\perp(x))$, center x and $(n - k)$-dimensional volume equal to $\lambda_{n-k}(A \cap E^\perp(x))$. Let

$$S_E(A) := \bigcup_{x \in \pi_E(A)} A_x.$$

The function $A \mapsto S_E(A)$ is a symmetrization (with respect to E). Obviously, for $k = n - 1$ it is the Steiner symmetrization. For $k = 1$ the symmetrization S_E is called the *Schwarz symmetrization*.

As well as the Steiner symmetrization, the Schwarz symmetrization preserves $\mathcal{K}^n$ and $\mathcal{K}^n_0$, preserves volume, and is not induced by any transformation of R^n (Exercises 4.3–4.5).

We end this section with one more well-known example of symmetrization with respect to a hyperplane.

4.3.2. DEFINITION. Let $H \in \mathcal{G}_{n-1}^n$. For every $A \in \mathcal{K}^n$ let

$$M_H(A) := \frac{1}{2}(A + \sigma_H(A)).$$

The function M_H is called the *Minkowski symmetrization* or the *Blaschke symmetrization* (compare [64]).

4.4 Means of rotations

We shall now consider the class of transformations of $\mathcal{K}^n$, which will be called means of rotations.[4]

4.4.1. DEFINITION. A map $T : \mathcal{C}^n \to \mathcal{C}^n$ is a *mean of rotations* if there exists $m \in \mathbb{N}$ and $f_1, \ldots, f_m \in SO(n)$ such that for every $A \in \mathcal{C}^n$

$$T(A) = \frac{1}{m} \sum_{i=1}^m f_i(A). \tag{4.11}$$

If T is defined by the formula (4.11), we say that it is *determined by* $f_1, \ldots, f_m$ or is the *mean of rotations* $f_1, \ldots, f_m$.

4.4.2. THEOREM. *Let* $T : \mathcal{C}^n \to \mathcal{C}^n$ *be a mean of rotations. Then*
(i) *T preserves the classes $\mathcal{K}^n$ and $\mathcal{K}_0^n$;*
(ii) *T preserves inclusion;*
(iii) *T is linear in the sense of Minkowski:*

$$T(\alpha A + \beta B) = \alpha T(A) + \beta T(B);$$

(iv) *the ball B^n is a fixed point of T;*
(v) *T is continuous (with respect to ϱ_H).*

Proof. (i) follows from 4.4.1 combined with 2.4.2, (ii) from 4.4.1 combined with 2.1.3, and (iii) from 4.4.1.

(iv): Let T be determined by $f_1, \ldots, f_m$. Since $f_i \in O(n)$ for every i, it follows that $f(B^n) = \frac{1}{m}(\sum_{i=1}^m B^n) = B^n$, by 2.3.6.

(v) is a consequence of (iii), (iv), and (1.4). $\square$

Means of rotations preserve mean width and do not increase diameter:

4.4.3. THEOREM. *Let* T *be a mean of rotations. For every* $A \in \mathcal{K}^n$,
(i) $\bar{b}(T(A)) = \bar{b}(A)$;
(ii) $\operatorname{diam} T(A) \leq \operatorname{diam} A$.

---

[4] In German *Drehmittelungen* ([30]). Hadwiger used this notion in a more general sense.

Proof. Let T be determined by $f_1, \ldots, f_m$. From Theorem 3.4.4, it follows directly that for every $u \in S^{n-1}$,

$$h(T(A), u) = \frac{1}{m} \sum_{i=1}^{m} h(f_i(A), u).$$

By the definition of the support function, 3.4.1,

$$h(f_i(A), u) = h(A, f_i^{-1}(u));$$

hence by 3.4.9, for every $u \in S^{n-1}$

$$b(T(A), u) = \frac{1}{m} \sum_{i=1}^{m} b(f_i(A), u) = \frac{1}{m} \sum_{i=1}^{m} b(A, f_i^{-1}(u)). \qquad (4.12)$$

Thus by 2.2.8, since the spherical measure σ is invariant under linear isometries, we obtain

$$\bar{b}(T(A)) = \frac{1}{\sigma(S^{n-1})} \int_{S^{n-1}} b(T(A), u)\, d\sigma(u)$$

$$= \frac{1}{m\sigma(S^{n-1})} \sum_{i=1}^{m} \int_{S^{n-1}} b(A, v)\, d\sigma(v) = \bar{b}(A).$$

From (4.12) and 2.2.7 it follows that

$$\mathrm{diam}\, T(A) = \sup_{v} b(T(A), v) \leq \mathrm{diam}\, A. \qquad \square$$

Means of rotations do not increase the values of the function r_0 (Definition 2.2.12).

4.4.4. THEOREM. *Let T be a mean of rotations. For every $A \in \mathcal{K}^n$,*

$$r_0(T(A)) \leq r_0(A).$$

Proof. In view of 4.4.2,

$$A \subset \alpha B^n \implies T(A) \subset \alpha B^n;$$

thus

$$\{\alpha > 0 \mid A \subset \alpha B^n\} \subset \{\alpha > 0 \mid T(A) \subset \alpha B^n\}.$$

This completes the proof. $\qquad \square$

The set of means of rotations is closed under superposition (Exercise 4.6):

4.4.5. PROPOSITION. *If T_1 and T_2 are means of rotations, then $T_1 T_2$ is also a mean of rotations.*

Generally, means of rotations are not induced by any transformations of R^n:

4.4.6. THEOREM. *Let T be a mean of rotations $f_1, \ldots, f_m$. Then $T = f^*$ for some $f : \mathbf{R}^n \to \mathbf{R}^n$ if and only if*

$$m = 1 \text{ or } f_1 = \cdots = f_m. \tag{4.13}$$

Proof. If condition (4.13) is satisfied, then $T = (f_1)^*$. Assume that it is not satisfied and let

$$T' := \left(\frac{1}{m} (f_1 + \cdots + f_m) \right)^*.$$

Then T' preserves the family of singletons, preserves inclusion, and $T'(A) \subset T(A)$ for every A. We shall show that

$$T' \neq T.$$

Indeed, let $A = B^n$. On the one hand, by 4.4.2 (v), $T(A) = A$; on the other hand, $T'(A)$ is the image of the unit ball under the map $\frac{1}{m}(f_1 + \cdots + f_m)$, which is a linear map but is not an isometry (Exercise 4.7), whence $T'(A) \neq A$.

Hence T is not induced by any transformation of $\mathbf{R}^n$ (compare with Theorem 4.1.5). $\qquad\square$

5
Rounding Theorems

Roughly speaking, rounding theorems[1] (5.1.2 and 5.3.2) show how to turn an arbitrary convex body into a ball, iterating Steiner symmetrizations or using means of rotations.

5.1 The first rounding theorem

5.1.1. DEFINITION. For any $A \in \mathcal{K}_0^n$, let

$$\mathbf{S}(A) := \{S_{H_k} \cdots S_{H_1}(A) \mid H_i \in \mathcal{G}_{n-1}^n \text{ for } i = 1, \ldots, k, \ k \in \mathbb{N}\}.$$

5.1.2. THEOREM. *For every $A \in \mathcal{K}_0^n$ there exists a sequence in $\mathbf{S}(A)$ that is Hausdorff convergent to a ball with center 0 and volume $V_n(A)$.*

Proof. Let $A \in \mathcal{K}_0^n$. We define

$$\alpha := \inf\{r_0(A') \mid A' \in \mathbf{S}(A)\}$$

(compare Definition 2.2.12). There exists a sequence $(A_i)_{i \in \mathbb{N}}$ in $\mathbf{S}(A)$ with

$$\alpha = \lim_i r_0(A_i).$$

For every i, the set A_i is obtained from A by iterating Steiner symmetrizations with respect to hyperplanes passing through 0. Since by 4.2.3 combined with 4.2.4 no Steiner symmetrization increases the value of r_0, it follows that

---

[1] *Kugelungstheoreme* in [29].

$$\forall i \; r_0(A_i) \leq r_0(A).$$

Thus the sequence $(A_i)_{i \in \mathbb{N}}$ is bounded, whence by 1.2.8 and 1.2.4, it has a convergent subsequence $(A_{k_i})_{i \in \mathbb{N}}$. Without loss of generality we may assume that $(A_i)_{i \in \mathbb{N}}$ is convergent; let

$$A_0 := \lim A_i. \tag{5.1}$$

Since the function r_0 is continuous (Theorem 2.2.13), it follows that

$$\alpha = r_0(A_0). \tag{5.2}$$

Let us notice that $\alpha > 0$. Indeed, by Theorem 4.2.8,

$$V_n(A_i) = V_n(A);$$

thus in view of 2.2.11,

$$V_n(A_0) \geq \limsup V_n(A_i) = V_n(A) > 0; \tag{5.3}$$

hence $\operatorname{int} A_0 \neq \emptyset$.

Let now $B_0 := \alpha B^n$. We shall prove that B_0 is the "snowball" we are looking for, i.e.,

$$A_0 = B_0. \tag{5.4}$$

By (5.2),

$$A_0 \subset B_0;$$

suppose that $A_0 \neq B_0$. Let $S_0 := \operatorname{bd} B_0$. Then $S_0 \setminus A_0$ is a nonempty open subset of S_0, whence there exists a "spherical ball" (the intersection of S_0 with a ball in $\mathbb{R}^n$) C_0 contained in S_0 and disjoint from A_0. By compactness of the sphere S_0, its covering by all congruent copies of C_0 has a finite subcovering $\{C_j \mid j = 0, \ldots, m\}$. For every $j \in \{1, \ldots, m\}$ there is a hyperplane $H_j \ni 0$ with $\sigma_{H_j}(C_{j-1}) = C_j$. It is easy to show (by induction on j) that

$$S_{H_m} \cdots S_{H_1}(A_0) \cap S_0 = \emptyset,$$

whence $S_{H_m} \cdots S_{H_0}(A_0) \subset \operatorname{int} B_0$. Since the symmetrizations S_{H_j} are continuous on $\mathcal{K}_0^n$ (Theorem 4.2.14), by (5.1) it follows that $S_{H_m} \cdots S_{H_1}(A_0) = \lim_i S_{H_m} \cdots S_{H_1}(A_i)$. Hence, for some i_0,

$$S_{H_m} \cdots S_{H_1}(A_{i_0}) \subset \operatorname{int} B_0.$$

Therefore, $r_0(S_{H_m} \cdots S_{H_1}(A_{i_0})) < \alpha$, though the set on the left-hand side belongs to $\mathbf{S}(A)$. This completes the proof of (5.4).

It remains to prove that

$$V_n(B_0) = V_n(A);$$

in view of (5.3), it suffices to prove the inequality

$$V_n(B_0) \leq V_n(A). \tag{5.5}$$

We apply 4.2.11 and 4.2.8. Since $A_i \in S(A)$ for every $i \in \mathbb{N}$, it follows that

$$\forall \varepsilon > 0 \; \forall i \; V_n((A)_\varepsilon) \geq V_n((A_i)_\varepsilon).$$

By (5.4), $B_0 = \lim A_i$; thus $B_0 \subset (A_i)_\varepsilon$ for almost all i, whence $V_n((A)_\varepsilon) \geq V_n(B_0)$ for every $\varepsilon > 0$. Passing to the lower bound with respect to ε, we obtain (5.5). $\qquad\square$

5.2 Applications of the first rounding theorem

As an application of Theorem 5.1.2, we shall give a proof of the *Brunn–Minkowski inequality*, which is the first part of the well-known Brunn–Minkowski Theorem (Theorem 5.2.1). The idea of this proof (for $n = 3$) is partially due to Wilhelm Blaschke (Blaschke applies Steiner symmetrizations, but does not use the rounding theorem; compare [5]). For a proof of the second part, concerning the equality case, see [64].

5.2.1. THEOREM. *Let $A_0, A_1 \in \mathcal{K}^n$. Then*
(i) *for every $t \in [0, 1]$,*

$$(V_n((1 - t)A_0 + tA_1))^{\frac{1}{n}} \geq (1 - t)(V_n(A_0))^{\frac{1}{n}} + t(V_n(A_1))^{\frac{1}{n}};$$

(ii) *the equality for some t holds if and only if A_0 and A_1 either are contained in parallel hyperplanes or are homothetic.*

Proof of (i). The inequality is evidently true for sets A_0, A_1 at least one of which has empty interior. Thus, let us assume that $A_0, A_1 \in \mathcal{K}_0^n$. For every $t \in [0, 1]$ let

$$A_t := (1 - t)A_0 + tA_1 \quad \text{and} \quad f(t) := (V_n(A_t))^{\frac{1}{n}}.$$

Then the inequality in Theorem 5.2.1 is equivalent to the condition

$$f(t) \geq (1 - t)f(0) + tf(1). \tag{5.6}$$

The inequality (5.6) is obviously true for $t = 0$ or $t = 1$; let $t \in (0; 1)$.

By Theorem 5.1.2, in the family $S(A_t)$ (Definition 5.1.1) there exists a sequence $(A_t^{(i)})_{i \in \mathbb{N}}$ Hausdorff convergent to the ball B_t with center 0 such that

$$V_n(B_t) = V_n(A_t), \tag{5.7}$$

and for every i, the set $A_t^{(i)}$ is obtained from A_t by means of a finite iteration $F_t^{(i)}$ of Steiner symmetrizations with respect to hyperplanes passing through 0:

$$A_t^{(i)} = F_t^{(i)}(A_t).$$

Let

$$\tilde{A}_0^{(i)} := F_t^{(i)}(A_0), \quad \tilde{A}_1^{(i)} := F_t^{(i)}(A_1). \tag{5.8}$$

By 4.2.3, 4.2.9, and 4.2.10, from (5.8) it follows that

$$A_t^{(i)} \supset (1-t)\tilde{A}_0^{(i)} + t\tilde{A}_1^{(i)}. \tag{5.9}$$

Two sequences defined by (5.8) are not necessarily convergent, but there exists an increasing sequence of indices $(k_i)_{i \in \mathbb{N}}$ such that $(\tilde{A}_0^{(k_i)})_{i \in \mathbb{N}}$ and $(\tilde{A}_1^{(k_i)})_{i \in \mathbb{N}}$ converge, respectively, to some $\tilde{A}_0$, $\tilde{A}_1$. From (5.9) it follows that

$$B_t \supset (1-t)\tilde{A}_0 + t\tilde{A}_1. \tag{5.10}$$

Further, by 5.1.2, there exists a sequence $(F_0^{(i)})_{i \in \mathbb{N}}$ of iterations of Steiner symmetrizations such that $\lim F_0^{(i)}(\tilde{A}_0) = B_0$; hence (5.10) implies

$$B_t \supset (1-t)B_0 + \hat{A}_1,$$

where $\hat{A}_1 = \lim_i F_0^{(l_i)}(\tilde{A}_1)$ for some subsequence (l_i). Applying now 5.1.2 to $\hat{A}_1$, we obtain

$$B_t \supset (1-t)B_0 + tB_1. \tag{5.11}$$

For every t, let r_t be the radius of B_t. From (5.11) we derive the inequality $r_t \geq (1-t)r_0 + tr_1$, and consequently, by (5.7), the corresponding inequality (5.6) for the nth roots of n-volumes. $\qquad\square$

5.2.2. COROLLARY. *For every* $A_0, \ldots, A_m \in \mathcal{K}^n$, $m \in \mathbb{N}$,

$$(V_n(A_0 + \cdots + A_m))^{\frac{1}{n}} \geq \sum_{i=0}^{m} (V_n(A_i))^{\frac{1}{n}}.$$

In view of 5.2.2, means of rotations do not decrease volumes of compact convex sets.

5.2.3. COROLLARY. *Let* $T : \mathcal{C}^n \to \mathcal{C}^n$ *be a mean of rotations. For every* $A \in \mathcal{K}^n$,

$$V_n(T(A)) \geq V_n(A).$$

5.3 The second rounding theorem

5.3.1. DEFINITION. For every $A \in \mathcal{K}_0^n$, let

$$\mathbf{T}(A) := \{T(A) \mid T \text{ is a mean of rotations}\}.$$

5.3.2. THEOREM. *Let* $A \in \mathcal{K}_0^n$ *and* $0 \in A$. *There exists a sequence in* $\mathbf{T}(A)$ *Hausdorff convergent to a ball* B_0 *satisfying the following conditions:*
(i) $\bar{b}(B_0) = \bar{b}(A)$;
(ii) $V_n(A) \leq V_n(B_0) \leq (\text{diam}A)^n \kappa_n$.

Proof. The reasoning is similar to that in the proof of Theorem 5.1.2.
Let $A \in \mathcal{K}_0^n$. Define

$$\alpha := \inf\{r_0(A') \mid A' \in \mathbf{T}(A)\}.$$

There is a sequence $(A_j)_{j \in \mathbb{N}}$ in $\mathbf{T}(A)$, such that

$$\alpha = \lim_j r_0(A_j).$$

Since for every j, the set A_j is obtained from A by a mean of rotations, from 4.4.2 it follows that

$$\forall j \ A_j \subset r_0(A)B^n.$$

Hence $(A_j)_{j \in \mathbb{N}}$ has a convergent subsequence, and thus we may assume that it is convergent; let

$$A_0 = \lim A_j.$$

By the continuity of the function r_0,

$$\alpha = r_0(A_0).$$

Let us notice that $\alpha > 0$. Indeed, by 5.2.3,

$$\forall j \ V_n(A_j) \geq V_n(A) > 0,$$

whence from Theorem 2.2.11 it follows that

$$V_n(A_0) \geq V_n(A) > 0; \tag{5.12}$$

thus $\text{int}(A_0) \neq \emptyset$.

Let $B_0 = \alpha B^n$. We shall prove that B_0 is the ball required, i.e.,

$$A_0 = B_0. \tag{5.13}$$

The inclusion $A_0 \subset B_0$ is evident. Suppose that $A_0 \neq B_0$. Let $S_0 = \text{bd}\,B_0$. Then $S_0 \setminus A_0$ is an open nonempty subset of the sphere S_0, whence there exists a "spherical ball" C_0 contained in $S_0 \setminus A_0$. Let $\{C_0, \ldots, C_m\}$ be a covering of S_0 by congruent copies of C_0. For every $i \in \{1, \ldots, m\}$ there exists $f_i \in SO(n)$ such that

$$f_i(C_0) = C_i.$$

For every $X \in \mathcal{K}^n$, let

$$T(X) := \frac{1}{m}(f_1(X) + \cdots + f_m(X)). \tag{5.14}$$

By the linearity of the support function (Theorem 3.4.4) and its geometric properties (Theorem 3.4.5),

$$\forall v \in S^{n-1} \; h(T(A_0), v) = \frac{1}{m} \sum_{i=1}^{m} h(f_i(A_0), v) = \frac{1}{m} \sum_{i=1}^{m} h(A_0, f_i^{-1}(v)).$$

$$(5.15)$$

In turn, since $A_0 \subset B_0$ and $h(B_0, u) = \alpha$ for every $u \in S^{n-1}$, it follows that

$$\forall v \in S^{n-1} \; \forall i \; h(A_0, f_i^{-1}(v)) \le \alpha. \tag{5.16}$$

Obviously, for every $v \in S^{n-1}$ there exists $i_0 \in \{1, \dots, m\}$ with $\alpha v \in C_{i_0}$. Thus $\alpha f_{i_0}^{-1}(v) = f_{i_0}^{-1}(\alpha v) \in C_0 \subset S_0 \setminus A_0$; hence

$$h(A_0, f_{i_0}^{-1}(v)) < \alpha. \tag{5.17}$$

From conditions (5.15)–(5.17) it follows that for every $v \in S^{n-1}$,

$$h(T(A_0), v) < \frac{1}{m} \sum_{i=1}^{m} \alpha = \alpha;$$

therefore, $\sup_v h(T(A_0), v) < \alpha$, because this upper bound is attained (the sphere is compact and the support function is continuous with respect to the second variable). However, $\sup_v h(T(A_0), v) = r_0(T(A_0))$, whence by 4.4.5 and the definition of r_0, for every j,

$$r_0(T(A_0)) < \alpha \le r_0(T(A_j)). \tag{5.18}$$

But T is continuous (Theorem 4.4.2 (v)), whence (5.18) contradicts the continuity of r_0. This proves condition (5.13).

It remains to prove (i) and (ii).

(i) is a consequence of continuity of the function $\bar{b}$ (see Theorem 2.2.10) and its invariance under means of rotations (Theorem 4.4.3 (i)).

(ii) The first inequality is a direct consequence of (5.12) and (5.13). Since $0 \in A$, it follows that $\alpha \le \operatorname{diam} A$. Hence

$$V_n(B_0) = \alpha^n \kappa_n \le (\operatorname{diam} A)^n \kappa_n. \qquad \square$$

5.4 Applications of the second rounding theorem

As an application of Theorem 5.3.2, we shall prove the following Bieberbach theorem:

5.4.1. THEOREM. *Among all the compact subsets of $\mathbf{R}^n$ with a given positive diameter, a ball has the maximal volume.*

Proof. We have to prove that if B is a ball and $A \in C^n$, then

$$\operatorname{diam} A = \operatorname{diam} B \implies V_n(A) \le V_n(B). \tag{5.19}$$

Since $V_n(A) \leq V_n(\text{conv} A)$ and $\text{diam} A = \text{diam}(\text{conv} A)$ (compare Exercise 3.7), we may assume that $A \in \mathcal{K}^n$.

The equality of diameters implies that $V_n(B) = (\frac{1}{2}\text{diam} A)^n \kappa_n$; thus it suffices to prove the inequality

$$V_n(A) \leq \frac{1}{2^n}(\text{diam} A)^n \kappa_n. \qquad (5.20)$$

By Theorem 5.3.2, there exists a sequence $(A_i)_{i \in \mathbb{N}}$ in $\mathbf{T}(A)$ convergent to a ball B_0 with

$$V_n(B_0) \geq V_n(A). \qquad (5.21)$$

Since means of rotations do not increase diameter (Theorem 4.4.3 (ii)), it follows that $\text{diam} A_i \leq \text{diam} A$ for every i, and thus by 2.2.10,

$$\text{diam} B_0 \leq \text{diam} A.$$

This condition together with (5.21) implies

$$\frac{V_n(B_0)}{(\text{diam} B_0)^n} \geq \frac{V_n(A)}{(\text{diam} A)^n},$$

which is equivalent to (5.20). $\square$

6

Convex Polytopes

The family of convex polytopes in R^n is a rich subfamily of $\mathcal{K}^n$, which forms a separate subject of research ([28], [44]).

We shall deal only with selected problems, in particular, with the role of convex polytopes in geometry of compact convex sets (Theorems 6.3.1 and 6.3.2), analogous to the role of arbitrary geometric polyhedra in the topology of compact subsets of R^n.

6.1 Polyhedra and their role in topology

A nonempty subset S of R^n is a *simplex* if

$$S = \mathrm{conv}\{a_0, \dots, a_k\}$$

for some affine independent set of points $\{a_0, \dots, a_k\}$. Such points $a_0, \dots, a_k$ are called the *vertices of S*. A simplex with vertices $a_0, \dots, a_k$ will be denoted by $\Delta(a_0, \dots, a_k)$.[1]

Let us note that vertices are the only points of a simplex that do not belong to the relative interior of any segment contained in this simplex. As a consequence, we obtain the following.

6.1.1. PROPOSITION. *Every simplex determines uniquely the set of its vertices:*

---

[1] Compare 3.1.

$$\Delta(a_0, \ldots, a_k) = \Delta(b_0, \ldots, b_l) \Longleftrightarrow k = l \ \text{and} \ \{a_0, \ldots, a_k\} = \{b_0, \ldots, b_l\}.$$

In view of 6.1.1, the *dimension of a simplex* can be defined as follows: for every $k \in \mathbf{N}$,

$$\dim \Delta(a_0, \ldots, a_k) := k;$$

hence the dimension of a simplex is equal to the dimension of $\text{aff}\{a_0, \ldots, a_k\}$.

For example, every segment is a simplex of dimension 1, and every triangle is a simplex of dimension 2.

It is convenient to extend the above definition of a k-dimensional simplex as follows: every singleton is a *simplex of dimension* 0; the empty set is the *simplex of dimension* -1.

To introduce the notion of a *face of a simplex*, let S be a simplex of dimension $k \geq 0$ and let $S^{(0)}$ be the set of its vertices; then every simplex with vertices in the set $S^{(0)}$ is a *face of S*; in particular, if $k = 0$, then S and $\emptyset$ are the only faces of S.

A face S' of S is said to be *proper* if $S' \neq S$.

Thus for instance, $\{a_0\}$ and $\{a_1\}$ are proper faces of the segment $\Delta(a_0, a_1)$; vertices (more precisely, singletons $\{a_0\}, \{a_1\}, \{a_2\}$) and sides are proper faces of the triangle $\Delta(a_0, a_1, a_2)$.

Let $S^{(i)}$ be the set of i-dimensional faces of a simplex S. We define the relation $\prec$ in the set $\bigcup_i S^{(i)}$:

$$S_1 \prec S_2 \Longleftrightarrow S_1 \ \text{is a face of} \ S_2.$$

6.1.2. DEFINITION. A set $\mathcal{T}$ of simplices is called a *simplicial complex* if
(i) $S_1, S_2 \in \mathcal{T} \Longrightarrow S_1 \cap S_2 \prec S_i \ \text{for} \ i = 1, 2,$
(ii) $S \in \mathcal{T} \ \text{and} \ S' \prec S \Longrightarrow S' \in \mathcal{T}.$

A set $P \subset \mathbf{R}^n$ is a (*geometric*) *polyhedron* if there exists a simplicial complex $\mathcal{T}$ such that
(iii) $P = \bigcup \mathcal{T}.$

We then say that $\mathcal{T}$ is a *triangulation of the polyhedron* P, and P is a *geometric realization of the complex* $\mathcal{T}$:

$$P = |\mathcal{T}|$$

(compare with Exercises 6.1–6.3).

The set of simplices in $\mathcal{T}$ of dimension i will be denoted by $\mathcal{T}^{(i)}$ and called the *i-dimensional skeleton of $\mathcal{T}$*. The *dimension of $\mathcal{T}$* is the maximum of the dimensions of its simplices:

$$\dim \mathcal{T} := \max\{\dim S \mid S \in \mathcal{T}\}.$$

Let us note that "polyhedron" is often understood as a topological polyhedron, i.e., a set homeomorphic to a geometric polyhedron. However, since we deal only with geometric polyhedra, we omit the adjective "geometric."

Polyhedra play an important role in the topology of $\mathbf{R}^n$; for instance, every compact subset can be (in some sense) approximated by a sequence of polyhedra (compare [14], Theorem 1.10.18).

To any polyhedron P in $\mathbb{R}^n$ an integer $\chi(P)$ is assigned, the *Euler–Poincaré characteristic of P*. The function χ is defined as follows. First, the characteristic of a simplicial complex is defined (we denote it by the same symbol χ, since it will not lead to confusion):

6.1.3. DEFINITION. For any simplicial complex $\mathcal{T}$, let $k_i(\mathcal{T})$ be the number of its i-dimensional simplices. Then

$$\chi(\mathcal{T}) := \sum_{i=0}^{\dim \mathcal{T}} (-1)^i k_i(\mathcal{T}).$$

6.1.4. THEOREM. *If the geometric realizations of complexes $\mathcal{T}$ and $\mathcal{T}'$ coincide, then*

$$\chi(\mathcal{T}) = \chi(\mathcal{T}')$$

(compare with Exercise 6.14).

In view of Theorem 6.1.4, the Euler–Poincaré characteristic of a polyhedron P can be defined by the formula

$$\chi(P) := \chi(\mathcal{T})$$

for any triangulation $\mathcal{T}$ of P.

6.1.5. EXAMPLE. Let S be an n-dimensional simplex. Then

$$\chi(S) = 1 \quad \text{and} \quad \chi(\mathrm{bd}\,S) = 1 - (-1)^n.$$

Indeed, it suffices to verify these two formulae for the simplest triangulations $\mathcal{T}$ and $\mathcal{T}'$ of the polyhedra S and $\mathrm{bd}\,S$, respectively: $\mathcal{T}$ consists of all the nonempty faces of S and $\mathcal{T}'$ consists of all the proper (nonempty) faces of S.

Evidently, $k_i(\mathcal{T}) = \binom{n+1}{i+1}$ for $i \in \{0, \ldots, n\}$, while $k_i(\mathcal{T}') = k_i(\mathcal{T})$ for $i < n$ and $k_n(\mathcal{T}') = 0$.

Hence

$$\chi(S) = \sum_{i=0}^{n}(-1)^i \binom{n+1}{i+1} = \sum_{j=1}^{n+1}(-1)^{j-1}\binom{n+1}{j} = 1;$$

therefore,

$$\chi(\mathrm{bd}\,S) = 1 - (-1)^n.$$

An important property of the Euler–Poincaré characteristic is its topological invariance (see [36], p. 242, for the proof):

6.1.6. THEOREM. *If polyhedra P_1 and P_2 are homeomorphic, then*

$$\chi(P_1) = \chi(P_2).$$

In view of 6.1.6, the function χ can be extended over the family of all topological polyhedra.

6.2 Convex polytopes

Convex polytopes may be defined as convex polyhedra; however, we prefer the traditional Definition 6.2.1, which is much simpler. In view of Theorem 6.2.4, these two definitions are equivalent.[2]

6.2.1. DEFINITION. A nonempty subset P of $\mathbb{R}^n$ is a *convex polytope* if there exists a finite $X \subset \mathbb{R}^n$ with $\operatorname{conv} X = P$.

The dimension of a convex polytope P is defined by the formula

$$\dim P := \dim \operatorname{aff} P. \tag{6.1}$$

Let $\mathcal{P}^n$ be the family of all convex polytopes in $\mathbb{R}^n$ and let

$$\mathcal{P}_0^n := \mathcal{P}^n \cap \mathcal{K}_0^n.$$

By 3.2.8, every convex polytope is compact, whence

6.2.2. $\mathcal{P}^n \subset \mathcal{K}^n$.

6.2.3. LEMMA. *If $P \in \mathcal{P}^n$ and H is a support hyperplane of P, then $H \cap P \in \mathcal{P}^n$. Moreover, for every finite X,*

$$P = \operatorname{conv} X \implies H \cap P = \operatorname{conv}(H \cap X).$$

Proof. Obviously, we may assume that P is not contained in H.

By 6.2.1, $P = \operatorname{conv} X$ for some finite set X. We shall prove that

$$H \cap P = \operatorname{conv}(H \cap X). \tag{6.2}$$

Since $H \cap P = H \cap \operatorname{conv} X \supset \operatorname{conv}(H \cap X)$, it remains to prove the inclusion $\subset$ in (6.2).

Let E^+ be the support half-space of P with boundary H. Assume that $x \in H \cap P$; then $x \in H$ and there exist $k \in \mathbb{N}$, $x_1, \ldots, x_k \in X$, and $t_1, \ldots, t_k > 0$ such that $\sum_{i=1}^k t_i = 1$ and $x = \sum_{i=1}^k t_i x_i$.

It suffices to prove that $x_i \in H$ for every $i \in \{1, \ldots, k\}$.

If $k \geq 2$, then by the Carathéodory Theorem 3.1.5, there exists a simplex S of dimension at least 1 with vertices in $\{x_1, \ldots, x_k\} \subset P$ such that $x \in \operatorname{relint} S$. Thus $S \subset E^+$, whence $x \in \operatorname{int} E^+$, contrary to the assumption that $x \in H$.

Hence $k = 1$, and thus $x_1 = x \in H$. $\qquad\square$

A nonempty subset F of a convex polytope P is called a *proper face of* P if $F \neq P$ and F is a support set of P:

$$\exists u \in S^{n-1} \quad F = P \cap H(P, u).$$

The polytope P is the *improper face* of itself.

---

[2] In the literature, the word "polytope" has different meanings; see, e.g., [12]

Hence, in view of Lemma 6.2.3, every face of a convex polytope is again a convex polytope.

Faces of dimension 0 are called *vertices*, faces of dimension 1 are called *edges*, and $(n - 1)$-dimensional faces are called *facets*.

The set of all proper faces of a convex polytope P is denoted by $\mathcal{F}(P)$, and the set of faces of dimension i by $\mathcal{F}^{(i)}(P)$.

We shall now prove

6.2.4. THEOREM. *For every subset P of R^n, the following conditions are equivalent*:
(i) $P \in \mathcal{P}^n$;
(ii) P *is a convex geometric polyhedron*.

Proof. (i) $\Longrightarrow$ (ii): We have to prove that for every $P \in \mathcal{P}^n$ there exists a triangulation $\mathcal{T}(P)$ of P.

If dim $P = 0$, then the complex $\mathcal{T}(P)$ consists of one vertex. Let dim $P = k \geq 1$ and assume that the assertion is true for the polytopes of dimension $k - 1$; in particular, we already have $\mathcal{T}(F)$ for every proper face F of P, so that $F' \prec F$ implies $\mathcal{T}(F') \subset \mathcal{T}(F)$.

We choose $p \in \text{relint}\, P$ and define $\mathcal{T}(P)$ as follows:

$$\mathcal{T}(P) := \bigcup_{F \in \mathcal{F}^{k-1}(P)} (\mathcal{T}(F) \cup \{\text{conv}(S \cup \{p\}) \mid S \in \mathcal{T}(F)\}).$$

Thus, $\mathcal{T}(P)$ is built of the given triangulations of all facets of P and all the k-dimensional simplices with one vertex p and remaining vertices in relbd P.

It is clear that $\mathcal{T}(P)$ is a triangulation of P. Hence P is a polyhedron. By the assumption, it is convex.

(ii) $\Longrightarrow$ (i): Assume now that P is a convex polyhedron. Let $\mathcal{T}$ be its triangulation. It suffices to prove that

$$P = \text{conv}|\mathcal{T}^{(0)}|. \tag{6.3}$$

Of course, $|\mathcal{T}^{(0)}| \subset P$, whence conv$|\mathcal{T}^{(0)}| \subset P$, because the set P is convex.

In turn, if $x \in P$, then there exists $S \in \mathcal{T}$ such that $x \in S$. Since the vertices of the simplex S belong to $|\mathcal{T}^{(0)}|$, it follows that $S \subset \text{conv}|\mathcal{T}^{(0)}|$. This completes the proof of (6.3). □

Every convex polytope is the convex hull of the set of its vertices:

6.2.5. THEOREM. *Let $\{a_0, \ldots, a_k\}$ be the set of vertices of a convex polytope P in R^n. Then*

$$P = \text{conv}\{a_0, \ldots, a_k\}.$$

Proof. The inclusion $\supset$ is obvious.

$\subset$: In view of Definition 6.2.1, the set P is the convex hull of a finite set X. Thus it suffices to prove that every vertex belongs to X. Suppose, to the contrary, that there exists a vertex a of P that does not belong to X. Then by the Carathéodory

Theorem 3.1.5, there exist an affine independent subset $\{x_0, \ldots, x_m\}$ of X and $t_0, \ldots, t_m$ such that

$$a = \sum_{i=0}^{m} t_i x_i, \ 0 \le t_i < 1, \ \sum_{i=0}^{m} t_i = 1.$$

Thus P contains a simplex S of positive dimension such that

$$a \in \text{relint} S,$$

contrary to the assumption that a is a vertex of P.   □

It is easy to prove that the class $\mathcal{P}^n$ is affine invariant (compare Exercise 6.6):

6.2.6. PROPOSITION. *For every affine automorphism f of $\mathbf{R}^n$,*

$$P \in \mathcal{P}^n \implies f(P) \in \mathcal{P}^n.$$

The class $\mathcal{P}^n$ is closed with respect to Minkowski addition:

6.2.7. PROPOSITION. $P, Q \in \mathcal{P}^n \implies P + Q \in \mathcal{P}^n$.
Proof. Let

$$P = \text{conv}\{a_0, \ldots, a_k\}, \ Q = \text{conv}\{b_0, \ldots, b_l\}.$$

By 3.2.6,

$$P + Q = \text{conv}\{a_i + b_j \mid i = 1, \ldots, k, \ j = 1, \ldots, l\}.$$ □

For any $P \in \mathcal{P}_0^n$ and any vector $v \ne 0$, let $P(v)$ be the support set of P with outer normal vector v (compare with 2.2.1) and let

$$\hat{P}(v) := \text{conv}(\{0\} \cup P(v)). \tag{6.4}$$

6.2.8. LEMMA. *Let $P, Q \in \mathcal{P}_0^n$ and $v \in S^{n-1}$. If at least one of the faces $P(v), Q(v)$ is of dimension $n - 1$, then $\dim(P + Q)(v) = n - 1$.*
Proof. By 2.2.3,
$$(P + Q)(v) = P(v) + Q(v).$$

Since $P(v)$ and $Q(v)$ are contained in hyperplanes parallel to $v^{\perp}$, it follows that $P(v) + Q(v)$ is also contained in such hyperplane; hence

$$\dim(P + Q)(v) \le n - 1.$$

But Minkowski addition does not decrease dimension (Exercise 6.7), whence

$$\dim(P + Q)(v) = n - 1.$$ □

The following theorem describes the Minkowski sum of convex polytopes in $\mathbf{R}^n$ with nonempty interiors.

6.2.9. THEOREM. *Let* $P, Q \in \mathcal{P}_0^n$, $0 \in \text{int}\, Q$, *and let* $\{v_1, \ldots, v_k\} \subset S^{n-1}$ *be the set of outer normal vectors to the facets of* P *or* Q. *Then*

$$P + Q = P \cup \bigcup_{i=1}^{k} (P(v_i) + \hat{Q}(v_i)),$$

where all the sets P, $P(v_i) + \hat{Q}(v_i)$, *and* $P(v_j) + \hat{Q}(v_j)$ *for* $i \neq j$ *have pairwise disjoint interiors.*

Proof. Evidently, the inclusion $\supset$ holds, and

$$P + Q = P \cup \bigcup_{0 < \lambda \leq 1} \text{bd}(P + \lambda Q).$$

Since for every $\lambda \in (0; 1]$ and $v \in S^{n-1}$,

$$\lambda \hat{Q}(v) \subset \hat{Q}(v)$$

(see (6.4)), to prove the inclusion $\subset$ it suffices to show that

$$\forall \lambda \in (0; 1] \;\; \text{bd}(P + \lambda Q) \subset \bigcup_{i=1}^{k} (P(v_i) + \lambda \hat{Q}(v_i)). \tag{6.5}$$

Let us note that (by Lemma 6.2.8) for any $P, Q' \in \mathcal{P}_0^n$, if $\{v_1, \ldots, v_k\}$ is the set of all outer normal unit vectors for the facets of P or Q', then

$$\text{bd}(P + Q') = \bigcup_{i=1}^{k} (P + Q')(v_i) = \bigcup_{i=1}^{k} (P(v_i) + Q'(v_i)) \subset P \cup \bigcup_{i=1}^{k} (P(v_i) + \hat{Q}'(v_i)).$$

Setting $Q' := \lambda Q$, we obtain (6.5).

It remains to prove that the summands on the right-hand side of (6.5) have pairwise disjoint interiors. We leave it to the reader (Exercise 6.8). $\qquad \square$

6.3 Approximation of convex bodies by polytopes

6.3.1. THEOREM (see [30]). *Let* $A \in \mathcal{K}^n$, $\varepsilon > 0$, *and let* X *be a polytope contained in* A. *Then there exists* $P \in \mathcal{P}^n$ *such that*

$$X \subset P \subset A \subset (P)_\varepsilon.$$

Proof. Since A is compact, its covering by all the balls with centers in A has a finite subcovering $\{B_1, \ldots, B_k\}$ such that $B_i = \{x_i\}_\varepsilon$ and all the vertices of X belong to $\{x_1, \ldots, x_k\}$. Let

$$P := \text{conv}\{x_1, \ldots, x_k\}.$$

Then
$$X \subset P \subset A,$$
because A is convex and by 6.2.5, the polytope X is the convex hull of the set of its vertices.

It remains to prove that $A \subset (P)_\varepsilon$.

Let $x \in A$. There exists $i \in \{1, \ldots, k\}$ such that $x \in B_i$, whence
$$\varrho(x, P) \leq \varrho(x, x_i) \leq \varepsilon.$$

Thus $x \in (P)_\varepsilon$. $\square$

6.3.2. THEOREM (compare [30]). *For every $A \in \mathcal{K}^n$ with $0 \in \mathrm{relint} A$ there exists a family of convex polytopes, $(P(\lambda))_{\lambda > 1}$, such that*
$$\forall \lambda > 1 \ \ P(\lambda) \subset A \subset \lambda \cdot P(\lambda)$$

and
$$\lim_{\lambda \to 1} \varrho_H(P(\lambda), A) = 0.$$

Proof. Since $\mathrm{aff} A$ is isometric to $\mathbf{R}^m$ for $m = \dim A$, without loss of generality we may assume that $A \in \mathcal{K}_0^n$.

Let X be an n-dimensional cube with center 0 contained in A, and let α be the length of its edge. For every $\lambda > 1$, let
$$\varepsilon(\lambda) := \frac{\alpha}{2}(\lambda - 1).$$

By Theorem 6.3.1, for any $\lambda > 1$ there is a convex polytope $P(\lambda)$ that satisfies the condition
$$X \subset P(\lambda) \subset A \subset (P(\lambda))_{\varepsilon(\lambda)}.$$

Since $\lim_{\lambda \to 1} \varepsilon(\lambda) = 0$, it follows that $\varrho_H(A, P(\lambda)) \to 0$ if $\lambda \to 1$.

It suffices to prove that for every λ,
$$(P(\lambda))_{\varepsilon(\lambda)} \subset \lambda \cdot P(\lambda). \tag{6.6}$$

Let $x \notin \lambda \cdot P(\lambda)$ for some λ; then there exists a point y with
$$\Delta(0, x) \cap \mathrm{bd}(\lambda \cdot P(\lambda)) = \{y\};$$
hence there exists a facet F of $P(\lambda)$ such that $y \in \lambda F$.

Let $H = \mathrm{aff} F$. Obviously, H is a support hyperplane of $P(\lambda)$. Let H' be the hyperplane parallel to H and passing through x. Then $H' \cap \lambda \cdot P(\lambda) = \emptyset$ and
$$\varrho(x, P(\lambda)) \geq \varrho(x, H) = \mathrm{dist}(H', H) > \mathrm{dist}(\lambda H, H) = \varrho(0, \lambda H) - \varrho(0, H)$$
$$= (\lambda - 1)\varrho(0, H) \geq (\lambda - 1)\varrho(0, \mathrm{bd} P) \geq (\lambda - 1)\varrho(0, \mathrm{bd} X)$$
$$\geq (\lambda - 1)\frac{\alpha}{2} = \varepsilon(\lambda).$$

Thus $x \notin (P(\lambda))_{\varepsilon(\lambda)}$. $\square$

6.4 Equivalence by dissection

For a given group G, the binary relation $\sim_G$ on the set $\mathcal{P}_0^n$ is defined as follows.

6.4.1. DEFINITION. Let G be a group of transformations of $\mathbf{R}^n$. Polytopes $P, Q \in \mathcal{P}_0^n$ are *equivalent by dissection with respect to G* (in symbols $P \sim_G Q$) if there exist $k \in \mathbf{N}$ and $P_1, \ldots, P_k, Q_1, \ldots, Q_k \in \mathcal{P}_0^n$ such that

$$P = \bigcup_{i=1}^{k} P_i, \quad Q = \bigcup_{i=1}^{k} Q_i, \quad P_i \equiv_G Q_i \text{ for } i = 1, \ldots, k,$$

and

$$\text{int}(P_i \cap P_j) = \emptyset = \text{int}(Q_i \cap Q_j) \text{ for } i \neq j.$$

Here G will be either the group of isometries or the group Tr of translations of $\mathbf{R}^n$. In the first case, the relation $\sim_G$ will be called simply *equivalence by dissection* and will be denoted by $\sim$.

6.4.2. THEOREM. *Let* $P, Q \in \mathcal{P}_0^2$. *Then*

$$P \sim Q \iff V_2(P) = V_2(Q).$$

This theorem was proved at the beginning of the nineteenth century (compare [7] and also [40] p. 246). The three-dimensional version of 6.4.2 was the subject of the third Hilbert problem: is it true that every two polytopes $P, Q \in \mathcal{P}_0^3$ with equal volumes are equivalent by dissection?[3]

The solution was given by Dehn a few months after Hilbert announced the problems (compare [57]). The following two three-dimensional simplices S_1, S_2 have equal volumes but are not equivalent by dissection:

$$S_1 = \text{conv}(A \cup \{(0, 0, 1)\}), \quad S_2 = \text{conv}(A \cup \{(0, 1, 1)\}),$$

where $A = \Delta((0, 0, 0), (1, 0, 0), (0, 1, 0))$.

To prove that S_1 is not equivalent to S_2, Dehn found invariants that distinguish one simplex from another. They are called Dehn's invariants.

Generally, a function $f : \mathcal{P}_0^n \to \mathbf{R}$ is a *Dehn invariant* if for every n-dimensional polytopes P_0, P_1, P_2,

$$P_1 \sim P_2 \Longrightarrow f(P_1) = f(P_2)$$

and

$$(P_0 = P_1 \cup P_2 \text{ and } \dim(P_1 \cap P_2) < n) \Longrightarrow f(P_0) = f(P_1) + f(P_2).$$

In the example considered, $f(S)$ is the sum of products of lengths of the edges of a three-dimensional simplex S by measures of the dihedral angles corresponding to these edges.

---

[3] See Section 8.6 of [39].

A complete classification of n-dimensional polytopes with respect to the relation $\sim_{Tr}$ is due to Hugo Hadwiger. Invariants that allow one to distinguish nonequivalent polytopes are called *Hadwiger's invariants* (see [57], p. 43).

As a consequence of Theorem 6.2.9 we obtain Theorem 6.4.4, the so-called *theorem on decomposition*; we shall need it only for $n \leq 3$. To formulate it, we extend in the obvious way the relation $\sim_{Tr}$ over finite unions of polytopes with disjoint interiors, and introduce the notion of cylindric polytope:

6.4.3. DEFINITION. A convex polytope $X \in \mathcal{P}_0^n$ is a *cylindric polytope* if there exist polytopes X_1, X_2 of positive dimensions such that X is their direct sum:

$$X = X_1 \oplus X_2;$$

i.e., $X = X_1 + X_2$, $\mathbb{R}^n = \text{aff}X_1 + \text{aff}X_2$, and $\text{aff}X_1 \cap \text{aff}X_2$ is a singleton.

6.4.4. THEOREM. *Let P, $Q \in \mathcal{P}_0^n$, $n \geq 2$, and $0 \in \text{int}Q$. There exist polytopes $Q_1, \ldots, Q_k \in \mathcal{P}_0^n$ and cylindric polytopes $X_1, \ldots, X_m$ such that*

$$P + Q = P \cup \bigcup_{i=1}^{k} Q_i \cup \bigcup_{j=1}^{m} X_j,$$

$\bigcup_{i=1}^{k} Q_i \sim_{Tr} Q$, *and all the sets on the right-hand side of the formula have pairwise disjoint interiors.*

Sketch of proof (see Figure 6.1). Let $w_1, \ldots, w_k$ be the outer normal unit vectors corresponding to the facets of Q. We dissect Q into $\hat{Q}(w_1), \ldots, \hat{Q}(w_k)$ (see (6.4)):

$$Q = \bigcup_{i=1}^{k} \hat{Q}(w_i).$$

Further, let $v_1, \ldots, v_l$ be among the outer unit normal vectors corresponding to the facets of P that do not belong to $\{w_1, \ldots, w_k\}$.

Then in view of 6.2.9,

$$P + Q = P \cup \bigcup_{i=1}^{k}(P(w_i) + \hat{Q}(w_i)) \cup \bigcup_{j=1}^{l}(P(v_j) + \hat{Q}(v_j)).$$

If $\dim P(w_i) = 0$, then $P(w_i) + \hat{Q}(w_i)$ is a translate of $\hat{Q}(w_i)$; otherwise, it is the union of some cylindric polytopes and a translate of $\hat{Q}(w_i)$.

By the assumption on $v_1, \ldots, v_l$, all of $Q(v_1), \ldots, Q(v_l)$ have dimensions less than $n - 1$. If $\dim Q(v_j) = 0$ (that is, $\dim \hat{Q}(v_j) = 1$), then $P(v_j) + \hat{Q}(v_j)$ has dimension n and is a cylindric polytope. If $\dim Q(v_j) > 0$, then $P(v_j) + \hat{Q}(v_j)$ is the union of some cylindric polytopes.

The details are left to the reader (see Exercise 6.9).   □

The following corollary can be derived from 6.4.2 (Exercise 6.11).

6.4.5. COROLLARY. *Every cylindric polytope in $\mathbb{R}^3$ is equivalent by dissection to a cube.*

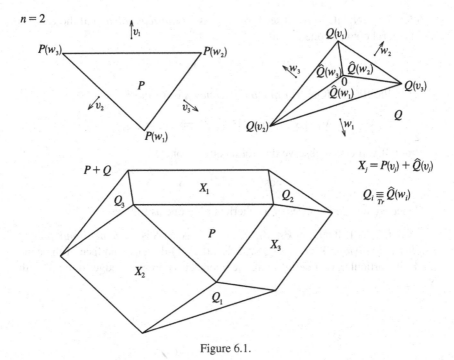

$$X_j = P(v_j) + \hat{Q}(v_j)$$

$$Q_i \underset{Tr}{\equiv} \hat{Q}(w_i)$$

Figure 6.1.

6.5 Spherical polytopes

We shall now introduce the notion of a spherical polytope, which will be useful in Chapter 7. It is closely related to that of a convex polytope in $\mathbf{R}^n$.

6.5.1. DEFINITION. A nonempty convex subset X of $\mathbf{R}^n$ will be called a *(linear) convex cone* if

$$\forall \lambda \geq 0 \; \lambda X \subset X$$

(see [64]).

It would be natural to refer to a translate by c of a linear convex cone as a *convex cone with vertex c*. However, we consider here only linear convex cones; thus we omit the adjective "linear."

6.5.2. PROPOSITION. *For every $A \in \mathcal{K}^n$, the set $\mathrm{pos}A$ is a convex cone.*

Proof is left to the reader (compare Exercise 6.12).

6.5.3. DEFINITION. A convex cone X in $\mathbf{R}^n$ is a *polyhedral convex cone* if there exist $u_1, \ldots, u_k \in S^{n-1}$ such that

$$X = \mathrm{conv}\left(\bigcup_{i=1}^{k} \mathrm{pos}u_i\right).$$

Any support set of the cone X is a *face*; a one-dimensional face is an *edge*.

6.5.4. DEFINITION. A subset Y of S^{n-1} is a *spherical polytope* if there exists a polyhedral convex cone X with

$$X \cap S^{n-1} = Y.$$

6.5.5. PROPOSITION. *For all convex cones X_1, X_2 in $\mathbb{R}^n$,*

$$X_1 \cap S^{n-1} = X_2 \cap S^{n-1} \implies X_1 = X_2.$$

Proof. It suffices to observe that for a convex cone X,

$$\mathrm{pos}(X \cap S^{n-1}) = X. \qquad \square$$

Hence we can define faces of a spherical polytope as follows:

6.5.6. DEFINITION. Let $k \in \{0, \ldots, n-1\}$. A set S is a *k-dimensional face of a spherical polytope Y in S^{n-1}* if $\mathrm{pos}\,S$ is a $(k+1)$-dimensional face of the cone $\mathrm{pos}\,Y$. In particular, *vertices of Y* are the intersection points of edges of $\mathrm{pos}\,Y$ with S^{n-1}.

7
Functionals on the Space $\mathcal{K}^n$.
The Steiner Theorem

7.1 Functionals on the space $\mathcal{K}^n$

In Chapters 2, 3, and 5, some functionals (that is, functions with real values) defined on $\mathcal{K}^n$ or $\mathcal{C}^n$, have already been considered. Before we return to those examples, let us specify the properties of the functionals in which we are interested.

7.1.1. DEFINITION. Let G be a group of transformations of $\mathbf{R}^n$. A functional $\Phi : \mathcal{K}^n \to \mathbf{R}$ is *invariant with respect to* G if for every $A, B \in \mathcal{K}^n$,

$$A \equiv_G B \implies \Phi(A) = \Phi(B).$$

If Φ is invariant with respect to the group of isometries of $\mathbf{R}^n$, we simply say that it is *invariant*.

7.1.2. DEFINITION. A functional $\Phi : \mathcal{K}^n \to \mathbf{R}$ is a *valuation* if for every $A, B \in \mathcal{K}^n$ with $A \cup B \in \mathcal{K}^n$,

$$\Phi(A \cup B) + \Phi(A \cap B) = \Phi(A) + \Phi(B). \tag{7.1}$$

The property described by (7.1) is sometimes called additivity (see [30]); however, to avoid a misunderstanding, we use this term only for Minkowski additivity. There is a connection between these two notions:

7.1.3. PROPOSITION. *Every Minkowski additive functional* $\Phi : \mathcal{K}^n \to \mathbf{R}$ *is a valuation.*

Proof. For every $A, B \in \mathcal{K}^n$ with convex union $A \cup B$,

$$(A \cup B) + (A \cap B) = A + B$$

(compare Exercise 2.1). Hence, from the additivity of Φ it follows that

$$\Phi(A) + \Phi(B) = \Phi(A + B) = \Phi((A \cup B) + (A \cap B))$$
$$= \Phi(A \cup B) + \Phi(A \cap B). \qquad \square$$

Generally, *continuity of* Φ will be understood as continuity with respect to the Hausdorff metric.

7.1.4. DEFINITION. A functional $\Phi : \mathcal{K}^n \rightarrow$ R is *increasing* if for every $A, B \in \mathcal{K}^n$,

$$A \subset B \Longrightarrow \Phi(A) \leq \Phi(B).$$

7.1.5. DEFINITION. Let $p \in$ R. A functional $\Phi : \mathcal{K}^n \rightarrow$ R is *homogeneous of degree p* if for every $A \in \mathcal{K}^n$ and $\lambda > 0$,

$$\Phi(\lambda A) = \lambda^p \Phi(A);$$

Φ is *homogeneous* if it is homogeneous of degree 1.

7.1.6. EXAMPLES. Minimal width, diameter, and mean width are continuous functionals (Theorem 2.2.10), as is the function r_0 (Theorem 2.2.13). The first three are invariant with respect to the group of all isometries, while the last one is invariant with respect to $O(n)$. Among these four functionals, only the mean width is a valuation (Exercise 7.1). All of them are increasing and homogeneous. $\qquad \square$

Obviously, the set of all functionals on $\mathcal{K}^n$, with addition and multiplication by scalars inherited from R^n, is a linear space.

The following theorem is a direct consequence of the above definitions.

7.1.7. THEOREM. *The set of functionals invariant with respect to a group G, the set of valuations, and the set of continuous functionals are linear subspaces of the space of all functionals on $\mathcal{K}^n$.*

The set of increasing functionals is closed under addition and multiplication by positive scalars.

All these considerations (Definitions 7.1.1, 7.1.2, 7.1.4, 7.1.5 and Theorem 7.1.7) can be restricted to any subset of $\mathcal{K}^n$, in particular to $\mathcal{P}^n$. Since nonempty compact convex sets can be approximated by convex polytopes (Theorem 6.3.2), a natural question is, when are the values of a functional on $\mathcal{K}^n$ uniquely determined by its values on $\mathcal{P}^n$? The answer is given by Theorem 7.1.11.

7.1.8. LEMMA. *If a functional $\Phi_0 : \mathcal{P}^n \rightarrow$ R is increasing, then for every $A \in \mathcal{K}^n$,*
(i) *the set $\{\Phi_0(P) \mid P \supset A\}$ is bounded from below;*
(ii) *the set $\{\Phi_0(P) \mid P \subset A\}$ is bounded from above.*

Proof. (i): Let $a \in A$. Then for every $P \in \mathcal{P}^n$,

$$P \supset A \Longrightarrow \Phi_0(P) \geq \Phi_0(\{a\}).$$

(ii): Let Q be an n-cube containing A. Then for every $P \in \mathcal{P}^n$

$$P \subset A \Longrightarrow \Phi_0(P) \leq \Phi_0(Q). \qquad \qquad \square$$

In view of Lemma 7.1.8, we may admit the following definition:

7.1.9. DEFINITION. For every increasing functional $\Phi_0 : \mathcal{P}^n \to \mathbb{R}$, let $\overline{\Phi}, \underline{\Phi} : \mathcal{K}^n \to \mathbb{R}$ be defined by the formulae

$$\overline{\Phi}(A) := \inf\{\Phi_0(P) \mid P \supset A, \ P \in \mathcal{P}^n\},$$

$$\underline{\Phi}(A) := \sup\{\Phi_0(P) \mid P \subset A, \ P \in \mathcal{P}^n\}.$$

7.1.10. LEMMA. If $\Phi_0 : \mathcal{P}^n \to \mathbb{R}$ is increasing, then $\overline{\Phi}$ and $\underline{\Phi}$ are also increasing and

$$\overline{\Phi}|\mathcal{P}^n = \Phi_0 = \underline{\Phi}|\mathcal{P}^n. \qquad (7.2)$$

If Φ_0 is homogeneous of degree p, then $\underline{\Phi}$ and $\overline{\Phi}$ are also homogeneous of degree p.

Proof. Let $A, B \in \mathcal{K}^n$ and let $A \subset B$. Then

$$\{\Phi_0(P) \mid P \supset B\} \subset \{\Phi_0(P) \mid P \supset A\},$$

whence $\overline{\Phi}(A) \leq \overline{\Phi}(B)$. Thus $\overline{\Phi}$ is increasing.

For $\underline{\Phi}$ the proof is analogous. Condition (7.2) is evident.

Now let Φ_0 be homogeneous of degree p. Then for every $\lambda > 0$,

$$\underline{\Phi}(\lambda A) = \sup_{P \subset \lambda A} \Phi_0(P) = \sup_{P' \subset A} \Phi_0(\lambda P') = \lambda^p \underline{\Phi}(A);$$

that is, $\underline{\Phi}$ is homogeneous of degree p. For $\overline{\Phi}$ the reasoning is analogous. $\square$

7.1.11. THEOREM. (i) If $\Phi_0 : \mathcal{P}^n \to \mathbb{R}$ is increasing, invariant with respect to Tr, and homogeneous of degree p, then $\underline{\Phi} = \overline{\Phi}$.

(ii) If, moreover, Φ_0 is uniformly continuous, then $\underline{\Phi}$ is continuous, whence it is a unique continuous extension of Φ_0 over $\mathcal{K}^n$.

Proof. (i): Let $A \in \mathcal{K}^n$. Since Φ_0 is increasing, it follows that for every $P, Q \in \mathcal{P}^n$,

$$P \subset A \subset Q \Longrightarrow \Phi_0(P) \leq \Phi_0(Q);$$

thus, passing first to $\sup_P$ and next to $\inf_Q$, in view of Definition 7.1.9 we obtain the inequality

$$\underline{\Phi}(A) \leq \overline{\Phi}(A).$$

Hence by Lemma 7.1.10,

$$P \subset A \subset Q \Longrightarrow \Phi_0(P) \leq \underline{\Phi}(A) \leq \overline{\Phi}(A) \leq \Phi_0(Q). \qquad (7.3)$$

Since Φ_0 is invariant with respect to Tr, we may assume that $0 \in \operatorname{relint} A$. Thus by Theorem 6.3.2, there exists a subfamily $(P(\lambda))_{\lambda > 1}$ of $\mathcal{P}^n$ such that

$$\forall \lambda > 1 \ P(\lambda) \subset A \subset \lambda P(\lambda) \tag{7.4}$$

and

$$\lim_{\lambda \to 1} \varrho_H(A, P(\lambda)) = 0. \tag{7.5}$$

From (7.3), (7.4), and homogeneity of degree p of Φ_0 it follows that for every $\lambda > 1$,

$$\Phi_0(P(\lambda)) \leq \underline{\Phi}(A) \leq \overline{\Phi}(A) \leq \Phi_0(\lambda P(\lambda)) = \lambda^p \Phi_0(P(\lambda)).$$

This condition together with (7.5) implies

$$\limsup_{\lambda} \Phi_0(P(\lambda)) = \underline{\Phi}(A) \leq \overline{\Phi}(A) \leq \limsup_{\lambda} \lambda^p \Phi_0(P(\lambda)) = \underline{\Phi}(A).$$

Therefore

$$\underline{\Phi}(A) = \overline{\Phi}(A). \tag{7.6}$$

By Lemma 7.1.10, the functional $\underline{\Phi}$ is increasing and homogeneous of degree p.

(ii): Assume Φ_0 to be uniformly continuous. Let $A_i \in \mathcal{K}^n$ for $i \in \mathbb{N}$ and $A = \lim_H A_i$. In view of Theorem 6.3.2, there exist sequences of polytopes $(P_k)_{k \in \mathbb{N}}$ and $(P_{i,k})_{k \in \mathbb{N}}$ for $i \in \mathbb{N}$ such that

$$A = \lim_H P_k \quad \text{and} \quad A_i = \lim_H P_{i,k}.$$

By the properties of limit, there exists an increasing sequence of indices $(k(i))_{i \in \mathbb{N}}$ such that

$$A = \lim_H P_{i,k(i)}.$$

Hence $\lim_i \varrho_H(P_{i,k(i)}, P_i) = 0$. Let $\varepsilon > 0$. Since Φ_0 is uniformly continuous, it follows that

$$\exists i_0 \forall i > i_0 \ |\Phi_0(P_{i,k(i)}) - \Phi_0(P_i)| < \frac{\varepsilon}{3}.$$

By 7.1.9 combined with (7.6),

$$\exists i_1 \forall i > i_1 |\Phi_0(P_{i,k(i)}) - \underline{\Phi}(A_i)| < \frac{\varepsilon}{3}$$

and

$$\exists i_2 \forall i > i_2 |\Phi_0(P_i) - \underline{\Phi}(A)| < \frac{\varepsilon}{3}.$$

Thus

$$\forall i > \max\{i_0, i_1, i_2\} \ |\underline{\Phi}(A) - \underline{\Phi}(A_i)| < \varepsilon.$$

We have proved that $\underline{\Phi}$ is continuous.

In view of 6.3.2, the set $\mathcal{P}^n$ is dense in $\mathcal{K}^n$, whence this continuous extension over $\mathcal{K}^n$ is unique.   □

The following proposition is sometimes called the "theorem on simultaneous approximation":

7.1.12. PROPOSITION. *If $\Phi_j : \mathcal{P}^n \to \mathbb{R}$, $j = 1, \ldots, m$, are increasing, then for every $\varepsilon > 0$ and $A \in \mathcal{K}^n$ there exist $P, Q \in \mathcal{P}^n$ such that*

$$P \subset A \subset Q, \quad |\Phi_j(P) - \underline{\Phi}_j(A)| \leq \varepsilon, \quad |\Phi_j(Q) - \overline{\Phi}_j(A)| \leq \varepsilon \quad \text{for } j = 1, \ldots, m.$$

Proof. Let $A \in \mathcal{K}^n$ and $\varepsilon > 0$. By Definition 7.1.9, for every $j \in \{1, \ldots, m\}$ there exist $P_j, Q_j \in \mathcal{P}^n$ such that

$$P_j \subset A \subset Q_j, \quad |\Phi_j(P_j) - \underline{\Phi}_j(A)| \leq \varepsilon, \quad |\Phi_j(Q_j) - \overline{\Phi}_j(A)| \leq \varepsilon.$$

Let

$$P := \operatorname{conv} \bigcup_{j=1}^{m} P_j, \quad Q := \bigcap_{j=1}^{m} Q_j.$$

Then $P \subset A \subset Q$,

$$\Phi_j(P) \leq \underline{\Phi}_j(A) \leq \Phi_j(P_j) + \varepsilon \leq \Phi_j(P) + \varepsilon,$$

and

$$\Phi_j(Q) \geq \overline{\Phi}_j(A) \geq \Phi_j(Q_j) - \varepsilon \geq \Phi_j(Q) - \varepsilon.$$

This completes the proof. $\qquad\qquad\qquad\qquad\qquad\qquad\qquad\qquad\square$

7.2 Basic functionals. The Steiner theorem

We are now going to define a finite sequence of functionals on $\mathcal{P}^n$ and extend them over $\mathcal{K}^n$. We begin with the notions of an outer normal angle and its measure.

7.2.1. DEFINITION. Let $P \in \mathcal{P}^n$, $k \in \{0, \ldots, n-1\}$, and $F \in \mathcal{F}^k(P)$. The set $\operatorname{nor}(P, F)$ and the real $\gamma(P, F)$ are defined as follows:

$$\operatorname{nor}(P, F) := \{u \in S^{n-1} \mid P(u) = F\},$$

$$\gamma(P, F) := \frac{\sigma_{n-k-1}(\operatorname{nor}(P, F))}{\omega_{n-k}}.$$

The set $\operatorname{nor}(P, F)$ is *the outer normal angle of the polytope P with respect to the face F*; the number $\gamma(P, F)$ is its (*normed*) *measure*.

7.2.2. EXAMPLE. (Figure 7.1). Let $\dim P = n$ and $F \in \mathcal{F}(P)$.

(a) If F is a facet of P with outer unit normal vector v, then $\operatorname{nor}(P, F) = \{v\}$ and $\gamma(P, F) = \frac{1}{2}$.

(b) If $\dim F = n - 2$, then F is the intersection of two facets F_1, F_2, and $\operatorname{nor}(P, F)$ is an arc of a great circle of the sphere S^{n-1}; this arc lies in the two-dimensional linear subspace orthogonal to F, and its endpoints are outer normals of F_1 and F_2.

(c) If $F = \{a\}$ (that is, F is a vertex of P), and $F_1, \ldots, F_m$ are the facets intersecting in a, then $\operatorname{nor}(P, F)$ is the spherical polytope whose vertices are outer normals of the facets $F_1, \ldots, F_m$. (Compare Definitions 6.5.4 and 6.5.6.)

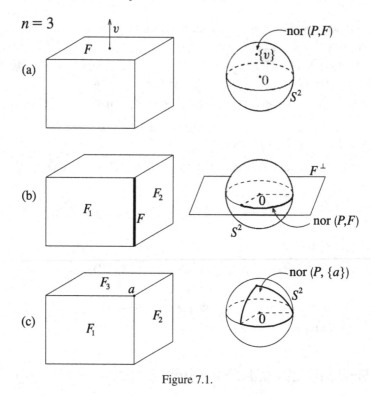

Figure 7.1.

7.2.3. EXAMPLE. Let $\dim P = n - 1$, $H := \operatorname{aff} P$, and let $v \in S^{n-1}$ be a normal vector of the hyperplane H. If $\dim F = n - 2$, then $\operatorname{nor}(P, F)$ is the semicircle with endpoints v and $-v$ in the two-dimensional linear subspace orthogonal to $\operatorname{aff} F$ (Figure 7.2). Hence $\gamma(P, F) = \frac{1}{2}$.

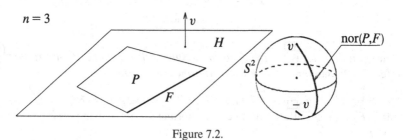

Figure 7.2.

We shall use the following convention: *the sum of numbers indexed by the empty set equals* 0.

7.2.4. DEFINITION. Let $P \in \mathcal{P}^n$ and $k \in \{0, \ldots, n - 1\}$.

$$V_k(P) := \sum_{F \in \mathcal{F}^k(P)} \lambda_k(F) \gamma(P, F).$$

The functionals $V_0, \ldots, V_n$ are called *basic functionals* or *intrinsic volumes*.

The term "basic functionals" is justified by Theorem 8.1.6.

The term "intrinsic volumes" was introduced by P. McMullen [45]. In earlier literature, instead of the basic functionals the so-called *Quermass integrals* $W_0, \ldots, W_n$ were considered. The relationship between $V_0, \ldots, V_n$ and $W_0, \ldots, W_n$ is described by the formula

$$W_k(P) := \binom{n}{k}^{-1} \kappa_k V_{n-k}(P) \tag{7.7}$$

(see [64]).

As we shall see in a while, (Theorem 7.2.5), unlike $W_0, \ldots, W_n$, basic functionals are independent of the dimension n of the space, and the symbol V_k for basic functional is compatible with the symbol V_n for n-dimensional volume, i.e., for n-dimensional Lebesgue measure:

7.2.5. THEOREM. *Let* $P \in \mathcal{P}^n$. *If* $\dim P \le k$, *then*

$$V_k(P) = \lambda_k(P).$$

Proof. Let $k \in \{0, \ldots, n-1\}$.

By Definition 7.2.4, if $\dim P < k$, then $V_k(P) = 0$, because $\mathcal{F}^k(P) = \emptyset$; thus $V_k(P) = \lambda_k(P)$.

Hence, we may assume that $\dim P = k$. Then P has a unique k-dimensional face:

$$\mathcal{F}^k(P) = \{P\},$$

nor(P, P) is an $(n - k - 1)$-dimensional subsphere of S^{n-1}, and $\gamma(P, P) = 1$. Thus $V_k(P) = \lambda_k(P)$, by Definition 7.2.4. $\qquad\square$

7.2.6. EXAMPLE. (a) If $P \in \mathcal{P}_0^n$, i.e., $\dim P = n$, then $V_{n-1}(P)$ equals one-half of the $(n-1)$-dimensional "surface area" of P:

$$V_{n-1}(P) = \frac{1}{2} \sum_{F \in \mathcal{F}^{n-1}(P)} \lambda_{n-1}(F).$$

This is a direct consequence of Definition 7.2.4 combined with Example 7.2.2 (a).

(b) From 7.2.4 and 7.2.2 (c), it follows that for every $P \in \mathcal{P}^n$,

$$V_0(P) = 1.$$

7.2.7. THEOREM. *The functional* $V_{n-1} : \mathcal{P}^n \to \mathbb{R}$ *is increasing.*

Proof. Let $P, Q \in \mathcal{P}_0^n$ and $P \subset Q$. Then every facet F of P is the image under orthogonal projection of a subset X_F of the boundary of Q; moreover, for different facets F_1, F_2 the set $X_{F_1} \cap X_{F_2}$ has dimension less than $n - 1$. Since orthogonal projection does not increase λ_{n-1}, by Example 7.2.6 (a) it easily follows that

$$V_{n-1}(P) \le V_{n-1}(Q).$$

We leave to the reader the proof for polytopes P, Q at least one of which has dimension less than n (Exercise 7.4). $\qquad\square$

Directly from the definitions of generalized ball and outer normal angle (1.1.5 and 7.2.1), we obtain the following theorem on the decomposition of α-hull of a polytope (Figure 7.3):

7.2.8. THEOREM. *For every $P \in \mathcal{P}^n$ and $\alpha > 0$,*

$$(P)_\alpha = P \cup \bigcup_{F \in \mathcal{F}} (F \oplus \alpha \cdot \mathrm{conv}(\{0\} \cup \mathrm{nor}(P, F))),$$

where all the summands of the union on the right-hand side of the equality have pairwise disjoint interiors, and the summands of the direct sum are contained in orthogonal affine subspaces.

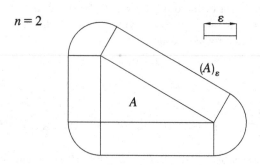

Figure 7.3.

We shall now prove the Steiner theorem for polytopes. It says that the volume of the α-hull of a polytope P is a polynomial in α whose coefficients depend only on $V_n(P), \ldots, V_0(P)$.

7.2.9. THEOREM. *For every $P \in \mathcal{P}^n$ and $\alpha > 0$,*

$$V_n((P)_\alpha) = \sum_{k=0}^{n} \alpha^{n-k} \kappa_{n-k} V_k(P).$$

Proof. By Theorem 7.2.8,

$$V_n((P)_\alpha) = V_n(P) + \sum_{k=0}^{n-1} \sum_{F \in \mathcal{F}^k(P)} \lambda_k(F) \cdot \lambda_{n-k}(\alpha \cdot \mathrm{conv}(\{0\} \cup \mathrm{nor}(P, F))).$$

$$(7.8)$$

Since λ_{n-k} is homogeneous of degree $n - k$ and

$$\frac{\lambda_{n-k}(\mathrm{conv}(\{0\} \cup \mathrm{nor}(P, F)))}{\kappa_{n-k}} = \frac{\sigma_{n-k-1}(\mathrm{nor}(P, F))}{\omega_{n-k}} = \gamma(P, F),$$

from (7.8) and Definition 7.2.4 it follows that

$$V_n((P)_\alpha) = V_n(P) + \sum_{k=0}^{n-1} \alpha^{n-k} \kappa_{n-k} \sum_{F \in \mathcal{F}^k} \lambda_k(F) \gamma(P, F)$$

$$= \sum_{k=0}^{n} \alpha^{n-k} \kappa_{n-k} V_k(P). \qquad \square$$

We shall now give a partial proof of the following theorem, which makes it possible to extend the basic functionals $V_0, \ldots, V_n$ over $\mathcal{K}^n$.

7.2.10. THEOREM. *For every $k \in \{0, \ldots, n\}$, the functional $V_k : \mathcal{P}^n \to \mathbb{R}$ is homogeneous of degree k, invariant, and increasing.*

Proof. Homogeneity of degree k and invariance follow directly from Definition 7.2.4.

We prove that V_k is increasing if $n \leq 3$.

Obviously, V_n is increasing. By Theorem 7.2.7 combined with Example 7.2.6 (b), V_{n-1} and V_0 are also increasing.

Let us show that V_1 is increasing: for every $P, Q \in \mathcal{P}^n$,

$$P \subset Q \Longrightarrow V_1(P) \leq V_1(Q). \qquad (7.9)$$

Indeed, for every $\alpha > 0$,

$$P \subset Q \Longrightarrow (P)_\alpha \subset (Q)_\alpha \Longrightarrow V_n((P)_\alpha) \leq V_n((Q)_\alpha),$$

and in view of the Steiner theorem for polytopes, 7.2.9, from the last inequality we infer that

$$\alpha^{n-1} \kappa_{n-1} V_1(P) + \sum_{k=2}^{n} \alpha^{n-k} \kappa_{n-k} V_k(P)$$

$$\leq \alpha^{n-1} \kappa_{n-1} V_1(Q) + \sum_{k=2}^{n} \alpha^{n-k} \kappa_{n-k} V_k(Q),$$

that is,

$$\alpha^{n-1} \kappa_{n-1}(V_1(Q) - V_1(P)) + \sum_{k=2}^{n} \alpha^{n-k} \kappa_{n-k}(V_k(Q) - V_k(P)) \geq 0.$$

Dividing both sides of the inequality by α^{n-1} and passing to the limit for $\alpha \to \infty$, we obtain

$$V_1(Q) - V_1(P) \geq 0,$$

which completes the proof of (7.9).

If $n \leq 3$, the proof is complete. If $n > 3$, it remains to show that $V_2, \ldots, V_{n-2}$ are increasing. We omit this part of proof, because it exceeds the scope of this book (compare with [64], p. 211). $\qquad \square$

The following corollary is a consequence of 7.2.10, 7.1.10, and 7.1.11 (i).

7.2.11. COROLLARY. *For every $k \in \{0, \ldots, n\}$, the functional $V_k : \mathcal{P}^n \to \mathrm{R}$ can be extended over $\mathcal{K}^n$ according to Definition 7.1.9. The extended functional is also homogeneous of degree k and increasing.*

To simplify the notation, we shall use the same symbols $V_0, \ldots, V_n$ for extended basic functionals.

We are now ready to prove the Steiner theorem for arbitrary compact convex sets, which is a generalization of 7.2.9.

7.2.12. THEOREM. *For every $A \in \mathcal{K}^n$ and $\alpha > 0$,*

$$V_n((A)_\alpha) = \sum_{k=0}^{n} \alpha^{n-k} \kappa_{n-k} V_k(A). \tag{7.10}$$

Proof. From the theorem on simultaneous approximation, 7.1.12, we deduce the existence of sequences $(P_i)_{i \in \mathrm{N}}$ and $(Q_i)_{i \in \mathrm{N}}$ in $\mathcal{P}^n$ such that $P_i \subset A \subset Q_i$ for every i and

$$\lim_i V_k(P_i) = V_k(A) = \lim_i V_k(Q_i) \quad \text{for} \quad k = 0, \ldots, n.$$

For these sequences

$$(P_i)_\alpha \subset (A)_\alpha \subset (Q_i)_\alpha,$$

whence

$$V_n((P_i)_\alpha) \le V_n((A)_\alpha) \le V_n((Q_i)_\alpha).$$

Applying the Steiner theorem for polytopes, 7.2.9, and passing to the limit for $i \to \infty$, we obtain the formula (7.10). $\qquad\square$

7.2.13. EXAMPLE. Let us calculate $V_k(B^n)$ for $k = 1, \ldots, n$.
For any $\alpha > 0$,

$$V_n((B^n)_\alpha) = (1+\alpha)^n \kappa_n = \sum_{k=0}^{n} \binom{n}{k} \alpha^{n-k} \kappa_n.$$

By Theorem 7.2.12, comparing the coefficients of α^{n-k}, we obtain

$$V_k(B^n) = \binom{n}{k} \frac{\kappa_n}{\kappa_{n-k}}.$$

Hence by (7.7),

$$W_{n-k}(B^n) = \kappa_n.$$

7.3 Consequences of the Steiner theorem

In view of the Steiner theorem, the n-dimensional volume $V_n((A)_\alpha)$ of the α-hull of a compact convex subset A of R^n is a polynomial in α whose coefficients are (up to some constant factors) intrinsic volumes of A.

Using this theorem we shall prove its analogue for the intrinsic volume $V_k((A)_\alpha)$ of the α-hull of A, where $k \in \{0, \ldots, n-1\}$.

7.3.1. THEOREM. *For every $A \in \mathcal{K}^n$, $\alpha > 0$, and $k \in \{0, \ldots, n\}$,*

$$V_k((A)_\alpha) = \frac{1}{\kappa_{n-k}} \sum_{i=0}^{k} \binom{n-i}{k-i} \alpha^{k-i} \kappa_{n-i} V_i(A). \tag{7.11}$$

Proof. Let us fix an $\alpha > 0$. For every $\beta > 0$,

$$(A)_{\alpha+\beta} = ((A)_\alpha)_\beta;$$

hence, applying the Steiner theorem 7.2.12 twice, we obtain for every $\beta > 0$ the following equality:

$$\sum_{i=0}^{n} (\alpha + \beta)^{n-i} \kappa_{n-i} V_i(A) = \sum_{k=0}^{n} (\beta^{n-k} \kappa_{n-k} V_k((A)_\alpha)). \tag{7.12}$$

Denote the left-hand side of (7.12) by L_β. Then

$$L_\beta = \sum_{i=0}^{n} \left(\sum_{j=0}^{n-i} \binom{n-i}{j} \alpha^j \beta^{n-(i+j)} \right) \kappa_{n-i} V_i(A).$$

For every $k \in \{0, \ldots, n\}$, the coefficients of β^{n-k} on both sides of (7.12) must be equal. In L_β, the $(n-k)$th power of β corresponds to $j = k - i$; thus

$$\sum_{i=0}^{k} \binom{n-i}{k-i} \alpha^{k-i} \kappa_{n-i} V_i(A) = \kappa_{n-k} V_k((A)_\alpha).$$

To obtain (7.11), it suffices to divide both sides of the last equality by κ_{n-k}.   □

We shall now make use of the Steiner theorem to examine properties of the basic functionals. For this purpose, we consider the family of functionals $\Phi_\alpha : \mathcal{K}^n \to \mathbb{R}$ for $\alpha > 0$:

$$\Phi_\alpha(A) := V_n((A)_\alpha). \tag{7.13}$$

Let us first prove the following.

7.3.2. THEOREM. (i) *For every $\alpha > 0$, the functional Φ_α is an invariant valuation;*

(ii) *For every $A_0 \in \mathcal{K}^n$ and every $\alpha_0 > 0$, the family $(\Phi_\alpha)_{\alpha \in (0, \alpha_0]}$ is equicontinuous in A_0.*

Proof. (i): Let $\alpha > 0$. Invariance of Φ_α (with respect to the isometries) is evident.

Let $A_1, A_2, A_1 \cup A_2 \in \mathcal{K}^n$. Let us note that

$$(A_1 \cup A_2)_\alpha = (A_1)_\alpha \cup (A_2)_\alpha \tag{7.14}$$

and

$$(A_1 \cap A_2)_\alpha = (A_1)_\alpha \cap (A_2)_\alpha. \tag{7.15}$$

Indeed,

$$(A_1 \cup A_2)_\alpha = \bigcup_{x \in A_1 \cup A_2} x + \alpha B^n = \bigcup_{x \in A_1} x + \alpha B^n \cup \bigcup_{x \in A_2} x + \alpha B^n = (A_1)_\alpha \cup (A_2)_\alpha.$$

Moreover, (7.14) holds for arbitrary nonempty closed subsets A_1, A_2 of $\mathbf{R}^n$.

In (7.15), the inclusion $\subset$ is evident (for arbitrary nonempty sets), and the inclusion $\supset$ follows directly from the condition

$$\varrho(x, A_1 \cap A_2) \leq \max_{i=1,2} \varrho(x, A_i),$$

which is true for arbitrary A_1, $A_2 \in \mathcal{K}^n$ with $A_1 \cup A_2 \in \mathcal{K}^n$ (Exercise 2.3).

Since V_n is a valuation, from (7.13)–(7.15) it follows that

$$\Phi_\alpha(A_1 \cup A_2) + \Phi_\alpha(A_1 \cap A_2) = \Phi_\alpha(A_1) + \Phi_\alpha(A_2);$$

that is, Φ_α is a valuation.

(ii): Let $A_0 \in \mathcal{K}^n$ and $\alpha_0 > 0$. We have to prove that for every $\varepsilon > 0$ there is a $\delta > 0$ such that

$$\forall \alpha \in (0, \alpha_0] \, \forall A \in \mathcal{K}^n \, \varrho_H(A, A_0) < \delta \implies |\Phi_\alpha(A) - \Phi_\alpha(A_0)| < \varepsilon. \quad (7.16)$$

Let $\varepsilon > 0$. Without loss of generality we may assume that $\varepsilon < 1$.

Let us first note that if $\varrho_H(A, A_0) < \varepsilon$, then there exists a ball B such that

$$(A)_{\alpha_0} \cup (A_0)_{\alpha_0} \subset B. \quad (7.17)$$

Indeed, $A \subset (A_0)_\varepsilon$, whence $(A)_{\alpha_0} \subset (A_0)_{\alpha_0 + \varepsilon} \subset B$ for some ball B.

Now let

$$\beta := \sum_{k=0}^{n-1} \kappa_{n-k} V_k(B), \quad \delta := \frac{\varepsilon}{\beta}. \quad (7.18)$$

Then $\beta > 1$, because from 7.2.6 (b) it follows that $V_0(B) = 1$. Hence $\delta < 1$.

Let $0 < \alpha \leq \alpha_0$ and $\varrho_H(A, A_0) < \delta$. Then

$$A_0 \subset (A)_\delta \quad \text{and} \quad A \subset (A_0)_\delta$$

and thus

$$(A_0)_\alpha \subset ((A)_\alpha)_\delta \quad \text{and} \quad (A)_\alpha \subset ((A_0)_\alpha)_\delta. \quad (7.19)$$

By the Steiner Theorem 7.2.12, from (7.19) it follows that

$$\Phi_\alpha(A_0) \leq \Phi_\alpha(A) + \sum_{k=0}^{n-1} \delta^{n-k} \kappa_{n-k} V_k((A)_\alpha)$$

and

$$\Phi_\alpha(A) \leq \Phi_\alpha(A_0) + \sum_{k=0}^{n-1} \delta^{n-k} \kappa_{n-k} V_k((A_0)_\alpha).$$

Applying now (7.17) and (7.18), we obtain

$$\Phi_\alpha(A_0) < \Phi_\alpha(A) + \delta\beta = \Phi_\alpha(A) + \varepsilon,$$

and analogously,

$$\Phi_\alpha(A) < \Phi_\alpha(A_0) + \varepsilon.$$

Hence

$$|\Phi_\alpha(A_0) - \Phi_\alpha(A)| < \varepsilon. \qquad \square$$

7.3.3. THEOREM. *The basic functionals are invariant and continuous valuations.*

Proof. By 7.3.2 (i), for every $\alpha > 0$, the function Φ_α defined by (7.13) is a valuation. Thus, by the Steiner theorem, for arbitrary $A_1, A_2 \in \mathcal{K}^n$ with $A_1 \cup A_2 \in \mathcal{K}^n$,

$$\sum_{k=0}^{n} \alpha^{n-k} \kappa_{n-k}(V_k(A_1 \cup A_2) + V_k(A_1 \cap A_2)) = \sum_{k=0}^{n} \alpha^{n-k} \kappa_{n-k}(V_k(A_1) + V_k(A_2)).$$

Comparing the coefficients of the corresponding powers of α, we obtain

$$V_k(A_1 \cup A_2) + V_k(A_1 \cap A_2) = V_k(A_1) + V_k(A_2)$$

for $k = 0, \ldots, n$. Thus $V_0, \ldots, V_n$ are valuations.

By 7.3.2 (i), the function Φ_α is invariant:

$$A \equiv B \implies \Phi_\alpha(A) = \Phi_\alpha(B).$$

Hence, applying again the Steiner theorem and comparing the coefficients of the corresponding powers of α, we obtain invariance of the basic functionals.

It remains to prove continuity.

Let $\lim_H A_i = A_0$ and $\alpha_0 > 0$. From 7.3.2 (ii) it follows that

$$\forall \alpha \in (0; \alpha_0] \lim_i \Phi_\alpha(A_i) = \Phi_\alpha(A_0). \qquad (7.20)$$

Let us take a sequence of distinct positive numbers $\alpha_1, \ldots, \alpha_{n+1} \in (0; \alpha_0]$.

Let $M(\alpha_1, \ldots, \alpha_{n+1})$ be the Vandermonde matrix;

$$M(\alpha_1, \ldots, \alpha_{n+1}) := \begin{pmatrix} 1 & \alpha_1 & (\alpha_1)^2 & \cdots & (\alpha_1)^n \\ 1 & \alpha_2 & (\alpha_2)^2 & \cdots & (\alpha_2)^n \\ \vdots & \vdots & \vdots & \ddots & \vdots \\ 1 & \alpha_{n+1} & (\alpha_{n+1})^2 & \cdots & (\alpha_{n+1})^n \end{pmatrix}$$

and let $f : R^{n+1} \to R^{n+1}$ be the linear transformation with the matrix $M(\alpha_1, \ldots, \alpha_{n+1})$ (in the canonical basis).

By the Steiner theorem, for $i \in \mathbb{N} \cup \{0\}$, $j = 1, \ldots, n+1$,

$$\Phi_{\alpha_j}(A_i) = \sum_{k=0}^{n}(\alpha_j)^{n-k}\kappa_{n-k}V_k(A_i).$$

Therefore, for a fixed i,

$$f(\kappa_0 V_n(A_i), \ldots, \kappa_n V_0(A_i)) = (\Phi_{\alpha_1}(A_i), \ldots, \Phi_{\alpha_{n+1}}(A_i)).$$

Since $M(\alpha_1, \ldots, \alpha_{n+1})$ is nonsingular ([50]), it follows that

$$f^{-1}(\Phi_{\alpha_1}(A_i), \ldots, \Phi_{\alpha_{n+1}}(A_i)) = (\kappa_0 V_n(A_i), \ldots, \kappa_n V_0(A_i)).$$

Hence from (7.20) it follows that

$$\lim_i V_k(A_i) = V_k(A_0)$$

for $k = 0, \ldots, n$. This proves the continuity of the basic functionals. □

We are now ready to prove the following.

7.3.4. PROPOSITION. $\sigma(S_{n-1}) = n \cdot \kappa_n$.

Proof. In view of 6.3.2 we can approximate the ball B^n by a sequence of convex polytopes $(P_k)_{k\in\mathbb{N}}$ contained in B^n. Of course, we may assume that the vertices of each P_k belong to S^{n-1}.

Let $\mathcal{T}_k$ be a triangulation of $\mathrm{bd}\,P_k$. Then

$$P_k = \bigcup_{\Delta \in \mathcal{T}_k} \mathrm{conv}(\{0\} \cup \Delta),$$

whence by continuity of V_n (see 7.3.3),

$$\kappa_n = V_n(B^n) = \lim_k V_n(P_k),$$

where $V_n(P_k) = \sum_{\Delta \in \mathcal{T}_k} \frac{1}{n}(V_{n-1}(\Delta) \cdot \varrho(0, \Delta))$. Thus $\kappa_n = \frac{1}{n}\sigma(S^{n-1})$. □

Let us observe that Proposition 7.3.4 can also be derived from the following well-known theorem on change of variables (compare with [66], formula (5.2.3) in Theorem 5.2.2).

7.3.5. THEOREM. *If f is a nonnegative measurable function on $\mathbb{R}^n$, then*

$$\int_{\mathbb{R}^n} f(x)d\lambda_n(x) = \int_{S^{n-1}} \int_0^\infty f(tu)t^{n-1}dtd\sigma(u).$$

Let $\rho_A : S^{n-1} \to R_+$ be the *radial function* of a convex body A in $\mathbb{R}^n$ with $0 \in A$:

$$\rho_A(u) := \sup\{\alpha \geq 0 \mid \alpha u \in A\}. \tag{7.21}$$

Setting $f = 1_A$, the characteristic function of A, as a direct consequence of Theorem 7.3.5 we obtain

7.3.6. COROLLARY. For every $A \in \mathcal{K}_0^n$,

$$V_n(A) = \frac{1}{n} \int_{S^{n-1}} \rho_A(u)^n d\sigma(u).$$

Of course, Prop. 7.3.4 is a particular case of Corollary 7.3.6, for $A = B^n$.

At the end of this section, we mention the traditional notation for $n = 3$ (compare [29]).

Let

$$V := V_3, \ F := 2V_2, \ M := \pi V_1, \ C := \frac{4}{3}\pi;$$

here V is the volume, F is the surface area (double for sets with empty interior), M is called the *mean curvature*, and C the *integral curvature*.

Let us notice that (in accordance with 7.2.6)

$$V = \kappa_0 V_3, \ F = \kappa_1 V_2, \ M = \kappa_2 V_1, \ C = \kappa_3 V_0. \tag{7.22}$$

Hence the Steiner formula for $n = 3$ can be rewritten as follows:

$$V((A)_\alpha) = V(A) + \alpha F(A) + \alpha^2 M(A) + \alpha^3 C(A).$$

8
The Hadwiger Theorems

8.1 The first Hadwiger theorem

From 7.1.7 and 7.3.3 it follows that every linear combination of basic functionals is a continuous invariant valuation. We shall now deal with the first Hadwiger theorem, which says that the family of continuous, invariant valuations coincides with the set of such linear combinations (Theorem 8.1.5).

Until 1995, the Hadwiger's original proof was the only one to be found in the literature. For the case $n = 3$, see [29]. The proof for arbitrary n was presented in [30]. It is very complicated and difficult to follow. In 1995, Daniel Klain gave a new proof of the Hadwiger theorem ([38]); however his methods go beyond the scope of this book.

In this situation, we present the proof of Theorem 8.1.5 only for $n = 3$; the proof for $n = 2$ is left to the reader (Exercise 8.1).

It is worthwhile to note that to generalize the proof for $n = 3$ to higher dimensions, it seems natural to generalize Lemmas 8.1.3 and 8.1.4. However, Theorem 6.4.5 on cylindric polytopes in $\mathbb{R}^3$, involved in the proof of Lemma 8.1.3 fails for $n > 3$ (compare with remarks following 6.4.2, and Exercise 6.13): some cylindric polytopes in $\mathbb{R}^n$ for $n > 3$ are not equivalent by dissection to an n-dimensional cube.

Lemmas 8.1.1–8.1.3 concern the so-called simple valuations. A valuation $\Psi : \mathcal{P}^n \to \mathbb{R}$ is *simple* (compare [38]) if for every $P \in \mathcal{P}^n$,

$$\dim P < n \implies \Psi(P) = 0.$$

8.1.1. LEMMA. *Let* $\Psi : \mathcal{P}^n \to \mathbb{R}$ *be a simple valuation. Then for every* $P_1, \ldots, P_m \in \mathcal{P}^n$ *with convex union* $\bigcup_{i=1}^m P_i$ *and pairwise disjoint interiors,*

$$\Psi\left(\bigcup_{i=1}^m P_i\right) = \sum_{i=1}^m \Psi(P_i). \tag{8.1}$$

Proof. Induction on m. For $m = 1$, equality (8.1) is an identity; for $m = 2$ it follows directly from the assumption. Let $m \geq 3$ and let the equality hold for $m - 1$. Obviously, we may assume that $\mathrm{int}\, P_i \neq \emptyset$ for $i = 1, \ldots, m$, because Ψ vanishes for polytopes with empty interiors.

Let $P := \bigcup_{i=1}^m P_i$. Since P is convex, there exist i_1, i_2 such that $\dim(P_{i_1} \cap P_{i_2}) = n - 1$ (Exercise 6.10). Let

$$H = \mathrm{aff}(P_{i_1} \cap P_{i_2}).$$

Then the hyperplane H dissects P into two convex polytopes P', P'':

$$P = P' \cup P'', \ P' \cap P'' = P \cap H.$$

Assume that $P_{i_1} \subset P'$, $P_{i_2} \subset P''$, and let

$$P_i' = P_i \cap P', \ P_i'' = P_i \cap P''.$$

Evidently, P_{i_1} and P_{i_2} are not dissected by H, whence

$$P_{i_1}' = P_{i_1}, \ P_{i_2}' = P_{i_2};$$

thus

$$P' = P_{i_1} \cup \bigcup_{i \neq i_1, i_2} P_i', \ P'' = P_{i_2} \cup \bigcup_{i \neq i_1, i_2} P_i''.$$

By the inductive assumption (since decompositions of P' and P'' have fewer than m elements),

$$\Psi(P') = \Psi(P_{i_1}) + \sum_{i \neq i_1, i_2} \Psi(P_i'), \ \Psi(P'') = \Psi(P_{i_2}) + \sum_{i \neq i_1, i_2} \Psi(P_i'');$$

hence

$$\Psi(P) = \Psi(P') + \Psi(P'')$$
$$= \Psi(P_{i_1}) + \Psi(P_{i_2}) + \sum_{i \neq i_1, i_2} (\Psi(P_i') + \Psi(P_i'')). \tag{8.2}$$

But P_i' and P_i'' have disjoint interiors and convex union P_i, whence

$$\Psi(P_i') + \Psi(P_i'') = \Psi(P_i) \ \text{for } i \neq i_1, i_2.$$

This condition combined with (8.2) implies (8.1). $\qquad\square$

The following lemma follows directly from 8.1.1.

8.1.2. LEMMA. *Let* $\Psi : \mathcal{P}^n \to \mathbb{R}$ *be an invariant simple valuation. Then, for arbitrary* $P \in \mathcal{P}^n$ *and* $Q_1, \ldots, Q_m \in \mathcal{P}^n$ *with pairwise disjoint interiors,*

$$P \sim \bigcup_{j=1}^{m} Q_j \implies \Psi(P) = \sum_{j=1}^{m} \Psi(Q_j).$$

8.1.3. LEMMA. *Let* $n \leq 3$ *and let* $\Psi : \mathcal{P}^n \to \mathbb{R}$ *be an invariant and continuous simple valuation satisfying the condition*
(i) $(\exists \alpha > 0 \ P \sim \alpha I^n) \implies \Psi(P) = 0.$
Then Ψ *is Minkowski linear.*

Proof. We have to prove that for every $m \in \mathbb{N}$,

$$\Psi\left(\sum_{i=1}^{m} t_i P_i\right) = \sum_{i=1}^{m} t_i \Psi(P_i).$$

Since, by 6.2.6 and 6.2.7, a linear combination of convex polytopes is a convex polytope, induction with respect to m is trivial. Thus it suffices to prove this condition for $m = 2$.

Since Ψ is a simple valuation, we may assume that $\dim P_i = n$ for $i = 1$ or $i = 2$; since Ψ is invariant, we may assume that $0 \in \mathrm{int}P_2$. In view of Theorem 6.4.4 on decomposition,

$$P_1 + P_2 = P_1 \cup \bigcup_k Q_k \cup \bigcup_j X_j,$$

where $\bigcup_k Q_k \sim_{Tr} P_2$, each of X_j is a cylindric polytope, and all the summands on the right-hand side of the equality have pairwise disjoint interiors.

Since Ψ is a simple valuation, by Lemma 8.1.1 it follows that

$$\Psi(P_1 + P_2) = \Psi(P_1) + \sum_k \Psi(Q_k) + \sum_j \Psi(X_j). \tag{8.3}$$

In turn, since $n \leq 3$, in view of 6.4.5 each X_j is equivalent by dissection to a cube, whence, by (i),

$$\forall j \ \Psi(X_j) = 0.$$

Hence by (8.3) combined with Lemma 8.1.2,

$$\Psi(P_1 + P_2) = \Psi(P_1) + \Psi(P_2); \tag{8.4}$$

thus Ψ is additive. It remains to prove that Ψ is homogeneous: for every $P \in \mathcal{P}^n$ and $t \in \mathbb{R}$,

$$\Psi(tP) = t\Psi(P). \tag{8.5}$$

For a fixed P, let us define $f : \mathbb{R} \to \mathbb{R}$ by the formula

$$f(t) := \Psi(tP).$$

Since multiplication by a scalar,

$$t \mapsto tP,$$

is continuous (Exercise 2.2) and, by the assumption, Ψ is continuous, it follows that f is continuous. In view of (8.4), the function f is additive:

$$f(t_1+t_2) = \Psi((t_1+t_2)P) = \Psi(t_1 P + t_2 P) = \Psi(t_1 P) + \Psi(t_2 P) = f(t_1) + f(t_2).$$

As is well known (compare [58] Theorem 10, p. 123), any continuous and additive real function is linear, and thus homogeneous; hence f is homogeneous. Therefore, $f(t) = tf(1)$; that is, (8.5) is satisfied.   □

8.1.4. LEMMA. *If $\Psi : \mathcal{K}^3 \to \mathbf{R}$ is an invariant and continuous simple valuation satisfying condition* (i) *of Lemma 8.1.3 for every $P \in \mathcal{P}^3$, then $\Psi = 0$.*

Proof. If P has empty interior, then $\Psi(P) = 0$.

Let $P \in \mathcal{P}_0^3$. Since Ψ is invariant, we may assume that $0 \in \mathrm{int}P$. By the second rounding theorem, 5.3.2, there exists a sequence $(A_i)_{i \in \mathbb{N}}$ in $\mathbf{T}(P)$ that is Hausdorff convergent to a ball B_0 with diameter $\bar{b}(P)$ and center 0. Let $i \in \mathbb{N}$. Since $A_i = \frac{1}{m} \sum_{j=1}^m g_j(P)$ for some $g_1, \dots, g_m \in O(n)$, by Lemma 8.1.3 for $n = 3$ it follows that

$$\Psi(A_i) = \frac{1}{m} \sum_{j=1}^m \Psi(g_j(P)),$$

where $\Psi(g_j(P)) = \Psi(P)$ because Ψ is invariant. Hence $\Psi(A_i) = \Psi(P)$ for every $i \in \mathbb{N}$, and thus by continuity,

$$\Psi(P) = \Psi(B_0). \tag{8.6}$$

Condition (8.6) holds for every $P \in \mathcal{P}_0^3$; thus in particular, it holds for $P = I^3$. Therefore, by (8.6), it follows that $\Psi(P) = 0$ for every polytope P.

Since Ψ is continuous, to complete the proof, it now remains to apply the approximation theorem 6.3.2.   □

8.1.5. THEOREM. *For every $\Phi : \mathcal{K}^n \to \mathbf{R}$ the following conditions are equivalent:*

(i) *Φ is an invariant and continuous valuation;*
(ii) *there exist $\alpha_0, \dots, \alpha_n \in \mathbf{R}$ such that*

$$\Phi = \sum_{i=0}^n \alpha_i V_i. \tag{8.7}$$

Proof. Implication (ii) $\Longrightarrow$ (i) follows from 7.1.7 combined with 7.3.3.

(i) $\Longrightarrow$ (ii) for $n = 3$: Let Φ be an invariant, continuous valuation. We are looking for $\alpha_0, \dots, \alpha_3$ such that the values of the functionals Φ and $\sum_{n=0}^3 \alpha_i V_i$

are the same for every $P \in \mathcal{P}^3$, and so in particular, for $P = \{0\}, I, I^2, I^3$, where I^k is a k-dimensional cube in $\mathbb{R}^3$, with 0 being a vertex ($I^0 := \{0\}$). Define $\alpha_0, \ldots, \alpha_3$:

$$\alpha_0 := \Phi(\{0\}), \quad \alpha_1 := (\Phi - \alpha_0 V_0)(I), \quad \alpha_2 := (\Phi - \alpha_0 V_0 - \alpha_1 V_1)(I^2),$$

$$\alpha_3 := (\Phi - \alpha_0 V_0 - \alpha_1 V_1 - \alpha_2 V_2)(I^3).$$

Let $\Psi : \mathcal{K}^3 \to \mathbb{R}$ be defined by the formula

$$\Psi := \Phi - \sum_{i=0}^{3} \alpha_i V_i.$$

It suffices to prove that $\Psi = 0$. To this end, let

$$\Phi_1 := \Phi - \alpha_0 V_0, \quad \Phi_2 = \Phi_1 - \alpha_1 V_1, \quad \Phi_3 := \Phi_2 - \alpha_2 V_2;$$

notice that

$$\Psi = \Phi_3 - \alpha_3 V_3. \tag{8.8}$$

Evidently, $\Phi_1, \Phi_2, \Phi_3, \Psi$ are invariant continuous valuations, (compare with 7.1.7) and

$$\alpha_i = \Phi_i(I^i) \text{ for } i = 1, 2, 3.$$

Since Φ is invariant and $V_0 = 1$ by 7.2.6 (b), it follows that

$$\Phi_1(\{a\}) = 0 \text{ for every } a \in \mathbb{R}^3.$$

Evidently, $\Phi_2(I) = 0$; we shall show that moreover,

$$\Phi_2(P) = 0 \text{ for every segment } P. \tag{8.9}$$

Since Φ_2 is invariant, it suffices to prove (8.9) for $P = t \cdot I$, where t is an arbitrary positive number. Let us note that the function $f : \mathbb{R} \to \mathbb{R}$ defined by the formula

$$f(t) := \Phi_2(t \cdot I)$$

is continuous, because Φ_2 is continuous, and is additive in view of Lemma 8.1.2 for $n = 1$; thus f is homogeneous. Hence

$$\Phi_2(P) = t\Phi_2(I) = 0,$$

which proves (8.9).

Now let P be a convex polygon in $\mathbb{R}^3$; then $\Phi_2(P)$ depends only on the area, $V_2(P)$. Indeed, by Theorem 6.4.2 and Lemma 8.1.2 for $n = 2$,

$$V_2(P') = V_2(P) \Longrightarrow P' \sim P \Longrightarrow \Phi_2(P) = \Phi_2(P').$$

Obviously, $\Phi_3(I^2) = 0$, whence $\Phi_3(P) = 0$ for every convex polygon P with area equal to 1. Let us show that moreover,

$$\Phi_3(P) = 0 \text{ for every convex polygon } P. \tag{8.10}$$

We define $g : \mathbb{R} \to \mathbb{R}$ by the formula

$$g(t) := \Phi_3(\sqrt{t} \cdot I^2).$$

Since $\Phi_3(P)$ depends only on the area of a polygon P, it follows that $g(t) = \Phi_3(P)$ for arbitrary polygon P with area $V_2(P) = t$. By Lemma 8.1.2 for $n = 2$, the function g is additive. Indeed, any square P with area $t_1 + t_2$ can be dissected into rectangles P_1, P_2 with areas t_1, t_2, respectively, and any rectangle is equivalent by dissection to a suitable square. Moreover, this function is continuous, whence it is homogeneous ([58], p. 123). Thus

$$\Phi_3(P) = t \cdot g(1) = t\Phi_3(I^2) = 0,$$

which proves (8.10).

From (8.8) and (8.10) it follows that Ψ vanishes for polygons, whence it is a simple valuation. Let us show that it satisfies condition (i) of Lemma 8.1.3. Of course, $\Psi(I^3) = 0$, and by the invariance of Ψ its value for any cube P depends only on $V_3(P)$.

Let

$$h(t) := \Psi(t^{\frac{1}{3}} I^3).$$

The function h is additive, because for $t = t_1 + t_2$ a cube P with volume t can be dissected into rectangular parallelepipeds P_1, P_2, with volumes t_1, t_2. These parallelepipeds are equivalent by dissection to suitable cubes. Since h is also continuous, it follows that h is homogeneous. Hence for every polytope P equivalent by dissection to a cube, with volume equal to t,

$$\Psi(P) = h(t) = th(1) = t\Psi(I^3) = 0.$$

Thus $\Psi = 0$ (by Lemma 8.1.4), which completes the proof. $\qquad\square$

In view of the Hadwiger Theorem 8.1.5, the linear space of invariant and continuous valuations on $\mathcal{K}^n$ is generated by the set $\{V_0, \ldots, V_n\}$ of the intrinsic volumes. Moreover, this set is a basis:

8.1.6. THEOREM. *The sequence* $(V_0, \ldots, V_n)$ *is a basis of the linear space of invariant continuous valuations on* $\mathcal{K}^n$.

Proof. In view of 8.1.5, it suffices to prove that the system of basic functionals is linearly independent:

$$\sum_{i=0}^{n} t_i V_i = 0 \implies t_i = 0 \text{ for } i = 0, \ldots, n. \tag{8.11}$$

Let us consider a sequence $(A_0, \ldots, A_n)$ in $\mathcal{K}^n$, with $\dim A_k = k$; for instance, let $A_k := \Delta(a_0, \ldots, a_k)$ for $k = 0, \ldots, n$. By the predecessor of the implication (8.11), for every $k \in \{0, \ldots, n\}$

$$\sum_{i=0}^{n} t_i V_i(A_k) = 0.$$

If $k = 0$, then $V_i(A_k) = 0$ for $i \geq 1$; since $V_0 = 1$ (Example 2.6 (b)), it follows that $t_0 = 0$.

Assume that $k > 0$ and $t_i = 0$ for $i \leq k - 1$. Then by the predecessor of (8.11),

$$\sum_{i=k}^{n} t_i V_i(A_k) = 0.$$

But in view of 7.2.5,

$$V_k(A_k) \neq 0 \text{ and } V_i(A_k) = 0 \text{ for every } i \in \{k + 1, \ldots, n\},$$

whence $t_k = 0$. This proves (8.11). $\qquad\square$

8.2 The second Hadwiger theorem

Since $V_k \geq 0$ for $k = 0, \ldots, n$, from 7.1.7 combined with 7.2.10 it follows that every linear combination of the basic functionals with nonnegative coefficients is an invariant increasing valuation.

The second Hadwiger theorem, 8.2.2, states that moreover, the class of invariant increasing valuations with nonnegative values coincides with the set of such linear combinations.

A proof of this theorem for $n = 3$ can be obtained by a suitable modification of the proof of 8.1.5 presented above (compare with Exercise 8.2). However, there is another possibility: this theorem can be derived from 8.1.5 and the following Mc-Mullen result (see Theorem 11.5 in [45]), which we cite without proof (compare with Exercise 8.3):

8.2.1. THEOREM. (P. McMullen) *If a valuation* $\Phi : \mathcal{K}^n \to R$ *is increasing and invariant with respect to the translations, then* Ψ *is continuous.*

8.2.2. THEOREM. *For every functional* $\Phi : \mathcal{K}^n \to R_+$ *the following conditions are equivalent:*
   (i) Φ *is an invariant and increasing valuation;*
   (ii) *there exist* $\alpha_0, \ldots, \alpha_n \in R_+$ *such that*

$$\Phi = \sum_{i=0}^{n} \alpha_i V_i.$$

9

Applications of the Hadwiger Theorems

9.1 Mean width and mean curvature

At the end of Chapter 7 we introduced the notion of mean curvature M for convex bodies in R^3. Generally, for $\mathcal{K}^n$, *mean curvature* is the functional M defined by the formula[1]

$$M := \frac{2\pi}{n-1} V_{n-2}. \tag{9.1}$$

The Hadwiger Theorem 8.1.5 yields a relationship between the mean width and the functional V_1 (Theorem 9.1.1), and thus for $n = 3$, between the mean width and mean curvature (Corollary 9.1.2).

9.1.1. THEOREM. *For every $A \in \mathcal{K}^n$,*

$$\bar{b}(A) = \frac{2\kappa_{n-1}}{n\kappa_n} V_1(A).$$

Proof. By 7.1.6, the mean width $\bar{b}$ is an invariant and continuous valuation. Hence in view of Theorem 8.1.5, there exist $\alpha_0, \ldots, \alpha_n \in R$ such that

$$\bar{b} = \sum_{i=0}^{n} \alpha_i V_i. \tag{9.2}$$

Obviously, for the ball $B := r \cdot B^n$ with arbitrary radius r,

---

[1] This functional is also called the *integral of mean curvature* (compare with [64], p. 210).

$$\bar{b}(B) = 2r;$$

since by 7.2.10, the functional V_i is homogeneous of degree i, from (9.2) it follows that

$$\forall r > 0 \; 2r = \sum_{i=0}^{n} \alpha_i V_i(B^n) \cdot r^i.$$

Thus $\alpha_i = 0$ for $i \neq 1$ and $\alpha_1 V_1(B^n) = 2$. Let us substitute $\alpha_0, \ldots, \alpha_n$ in (9.2); we obtain the formula

$$\bar{b} = \frac{2}{V_1(B^n)} V_1,$$

which, combined with 7.2.13, implies the required relationship between $\bar{b}$ and V_1. $\qquad\Box$

Directly from 9.1.1 and (9.1) we deduce

9.1.2. COROLLARY. *For every* $A \in \mathcal{K}^3$,

$$\bar{b}(A) = \frac{1}{2\pi} M(A).$$

The experienced reader will certainly appreciate Theorem 9.1.1 and Corollary 9.1.2. While it usually is difficult to calculate the mean width of a convex set A directly from Definition 2.2.8 (even if this set is simple and regular), it is much easier to find $V_1(A)$ (i.e., if $n = 3$, the mean curvature of A) and apply Theorem 9.1.1 or Corollary 9.1.2.

9.2 The Crofton formulae

The Crofton formulae (Theorem 9.2.6) are one of the most important and most interesting applications of the Hadwiger theorems. These integral formulae express, for $k \in \{1, \ldots, n\}$, the value of V_k at any $A \in \mathcal{K}^n$ by means of the values of V_{k-1} at sections of A by affine subspaces of some dimension $i \leq k - 1$.

Let us observe that our consideration may be restricted to $i = n - 1$, that is, to sections by hyperplanes. Indeed, every section of A by a subspace of dimension i for some $i \geq 1$ can be obtained as the result of $n - i$ operations of cutting a compact convex set contained in a subspace of dimension j by a subspace of dimension $j - 1$, for $j = n, \ldots, i + 1$.

Hence, let us consider the family $\mathcal{E}^n$ of hyperplanes in $\mathbf{R}^n$. In Section 2.5 we used the parametric representation $\phi : S^{n-1} \times \mathbf{R}_+ \to \mathcal{E}^n$ (see (2.7)) to define the limit in $\mathcal{E}^n$ (Definition 2.5.2). Thus we introduced a topology in $\mathcal{E}^n$.

A measure μ on the family $\mathcal{B}(\mathcal{E}^n)$ of Borel sets in $\mathcal{E}^n$ is defined as follows.

9.2.1. DEFINITION. Let $\mu_0 : \mathcal{B}(S^{n-1} \times \mathbf{R}) \to \mathbf{R}$ be the product measure of σ and λ_1.

The function $\mu : \mathcal{B}(\mathcal{E}^n) \to \mathbf{R}$ is defined by the formula

$$\forall \mathcal{X} \in \mathcal{B}(\mathcal{E}^n) \quad \mu(\mathcal{X}) := \mu_0(\phi^{-1}(\mathcal{X})).$$

9.2.2. PROPOSITION. *The function μ is a measure.*

(Compare with Exercise 9.2.)

9.2.3. EXAMPLES. (a) Fix a $v_0 \in S^2$ and $\alpha > 0$. Let

$$\mathcal{X} := \{E \in \mathcal{E}^3 \mid \mathrm{dist}(0, E) \in [0, \alpha], \ v_0 \perp E\}.$$

The set $\mathcal{X}$ is closed in $\mathcal{E}^3$, so it is a Borel set; since

$$\phi^{-1}(\mathcal{X}) = \{v_0\} \times [0, \alpha] \cup \{(-v_0, 0)\},$$

it follows that $\mu(\mathcal{X}) = \mu_0(\phi^{-1}(\mathcal{X})) = 0 \cdot \alpha = 0$.

(b) Let $A \in \mathcal{K}_0^n$, $0 \in A$ and let

$$\mathcal{X} = \{H(A, u) \mid u \in S^{n-1}\}.$$

Then by 3.4.5,

$$\phi^{-1}(\mathcal{X}) = \{(u, h_A(u)) \mid u \in S^{n-1}\} = \mathrm{graph}\, h_A;$$

thus $\phi^{-1}(\mathcal{X})$ is measurable as the graph of the continuous function $h_A|S^{n-1}$, and by Fubini's theorem, $\mu(\mathcal{X}) = \mu_0(\phi^{-1}(\mathcal{X})) = 0$.

For any $\Phi : \mathcal{K}^n \to \mathrm{R}$ and $A \in \mathcal{K}^n$ we define the function $\Phi_A : \mathcal{E}^n \to \mathrm{R}$ by the formula

$$\Phi_A(E) := \begin{cases} \Phi(A \cap E) & \text{if } A \cap E \neq \emptyset \\ 0 & \text{if } A \cap E = \emptyset. \end{cases} \tag{9.3}$$

The following statement is a direct consequence of 2.5.6 combined with 9.2.3 (b).

9.2.4. PROPOSITION. *If a functional $\Phi : \mathcal{K}^n \to \mathrm{R}$ is continuous, then for every $A \in \mathcal{K}^n$ the function Φ_A is continuous μ-almost everywhere, and thus μ-integrable.*

To every continuous functional $\Phi : \mathcal{K}^n \to \mathrm{R}$ we now assign the functional $\hat{\Phi} : \mathcal{K}^n \to \mathrm{R}$ defined by

$$\hat{\Phi}(A) := \int_{\mathcal{E}^n} \Phi_A(E) d\mu(E). \tag{9.4}$$

Let us show that the assignment $\Phi \mapsto \hat{\Phi}$ preserves the properties considered in the previous chapter.

For every $A \in \mathcal{K}^n$, let

$$\mathcal{E}_A := \{E \in \mathcal{E}^n \mid A \cap E \neq \emptyset\}. \tag{9.5}$$

9.2.5. THEOREM. *For every continuous $\Phi : \mathcal{K}^n \to \mathrm{R}$,*

(i) $\hat{\Phi}$ *is continuous;*

(ii) *if Φ is a valuation, then also $\hat{\Phi}$ is a valuation;*

(iii) *if Φ is invariant, then also $\hat{\Phi}$ is invariant.*

Proof. By (9.3)–(9.5), for every $A \in \mathcal{K}^n$,

$$\hat{\Phi}(A) = \int_{\mathcal{E}_A} \Phi(A \cap E)d\mu(E).$$

Condition (i) follows from Theorem 2.5.7 combined with Example 9.2.3 (b), since a hyperplane disjoint from the interior of A is either disjoint from A or is a support hyperplane of A.

Condition (ii) follows from the additivity of integral and definition of valuation.

(iii): Assume Φ to be invariant. Let $f : \mathbb{R}^n \to \mathbb{R}^n$ be an isometry. Then

$$\hat{\Phi}(f(A)) = \int_{\mathcal{E}_{f(A)}} \Phi(f(A) \cap E')d\mu(E') = \int_{f^{-1}(\mathcal{E}_{f(A)})} \Phi(f(A) \cap f(E))d\mu(E)$$

$$= \int_{\mathcal{E}_A} \Phi(f(A \cap E))d\mu(E) = \int_{\mathcal{E}_A} \Phi(A \cap E)d\mu(E) = \hat{\Phi}(A). \qquad \square$$

9.2.6. THEOREM. *Let $n \geq 2$. For $k \in \{1, \ldots, n\}$, let*

$$\beta_{n,k} := \frac{2}{k} \cdot \frac{\kappa_{k-1}}{\kappa_k \kappa_{n-1}}.$$

Then for every $A \in \mathcal{K}^n$,

$$V_k(A) = \beta_{n,k} \int_{\mathcal{E}_A} V_{k-1}(A \cap E)d\mu(E). \qquad (9.6)$$

Proof. Let $\Phi := V_{k-1}$. Then by (9.3)–(9.5), for every A,

$$\hat{\Phi}(A) = \int_{\mathcal{E}_A} V_{k-1}(A \cap E)d\mu(E). \qquad (9.7)$$

Since, by Theorem 7.3.3, the functional V_{k-1} is a continuous, invariant valuation, in view of Theorem 9.2.5, so is $\hat{\Phi}$. By the Hadwiger Theorem 8.1.5, there exist $\alpha_0, \ldots, \alpha_n$ such that

$$\hat{\Phi} = \sum_{i=0}^{n} \alpha_i V_i. \qquad (9.8)$$

To find $\alpha_0, \ldots, \alpha_n$, let us calculate $\hat{\Phi}(r B^n)$ for arbitrary positive r.

Every hyperplane E is of the form $\phi(v, t)$, where $v \perp E$ and $t = \text{dist}(0, E)$ (compare (2.7) and 2.5.1); thus by definition of the measure μ (Definition 9.2.1) combined with (9.7),

$$\hat{\Phi}(r B^n) = \int_{S^{n-1}} \int_0^r V_{k-1}(r B^n \cap \phi(v, t))dt d\sigma(v).$$

Since for $t \in (0; r)$ the intersection of $r B^n$ and $\phi(v, t)$ is an $(n - 1)$-dimensional ball with radius $\sqrt{r^2 - t^2}$, and V_{k-1} is homogeneous of degree $k - 1$ (Theorem 7.2.11), it follows that

$$\hat{\Phi}(rB^n) = \sigma(S^{n-1})V_{k-1}(B^{n-1})\int_0^r (\sqrt{r^2-t^2})^{k-1}dt.$$

By substitution $t = r\sin s$ for $s \in [0, \frac{\pi}{2}]$, we obtain

$$\hat{\Phi}(rB^n) = \sigma(S^{n-1})V_{k-1}(B^{n-1})\left(\int_0^{\frac{\pi}{2}} \cos^k s\, ds\right)r^k.$$

But, by Fubini's theorem,

$$\kappa_k = V_k(B^k) = 2\int_0^1 (\sqrt{1-t^2})^{k-1}V_{k-1}(B^{k-1})dt = 2\kappa_{k-1}\int_0^{\frac{\pi}{2}} \cos^k s\, ds$$

(Figure 9.1), whence

$$\int_0^{\frac{\pi}{2}} \cos^k s\, ds = \frac{\kappa_k}{2\kappa_{k-1}}.$$

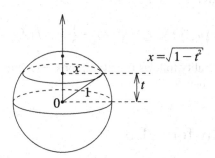

Figure 9.1.

Thus

$$\hat{\Phi}(rB^n) = \sigma(S^{n-1})V_{k-1}(B^{n-1})\frac{\kappa_k}{2\kappa_{k-1}}\cdot r^k.$$

By this formula together with (9.8), comparing the coefficients of r^i, we obtain $\alpha_i = 0$ for $i \neq k$ and

$$\alpha_k = \sigma(S^{n-1})\frac{V_{k-1}(B^{n-1})}{V_k(B^n)}\frac{\kappa_k}{2\kappa_{k-1}}. \qquad (9.9)$$

Since by 7.3.4,

$$\sigma(S^{n-1}) = n\kappa_n,$$

from 7.2.13 combined with (9.9) it follows that

$$\alpha_k = \frac{k\kappa_{n-1}\kappa_k}{2\kappa_{k-1}}.$$

It now remains to insert $\alpha_0, \ldots, \alpha_n$ into (9.8). $\qquad\square$

Using the traditional notation for $n = 3$ (see formulae (7.22) at the end of Chapter 7), we can rewrite the Crofton formulae for this particular case as follows.

9.2.7. THEOREM. *Let $A \in \mathcal{K}^3$. Then*

$$V(A) = \frac{1}{4\pi} \int_{\mathcal{E}_A} F(A \cap E) d\mu(E),$$

$$F(A) = \frac{4}{\pi^3} \int_{\mathcal{E}_A} M(A \cap E) d\mu(E),$$

$$M(A) = \frac{1}{4\pi} \int_{\mathcal{E}_A} C(A \cap E) d\mu(E)$$

(Exercise 9.4).   □

From the Crofton theorem we derive the following simple relationship between the measure of $\mathcal{E}_A$ and the mean width of A:

9.2.8. COROLLARY. *For every $A \in \mathcal{K}^n$,*

$$\mu(\mathcal{E}_A) = \kappa_{n-1} V_1(A) = \frac{1}{2} n \kappa_n \cdot \bar{b}(A).$$

Proof. The first equality follows from (9.6) for $k = 1$; applying Theorem 9.1.1 we obtain the second one.   □

9.3 The Cauchy formulae

The so-called Cauchy formulae (Theorem 9.3.2) are another important application of the Hadwiger theorems. These integral formulae express, for $k \in \{1, \ldots, n\}$, the value of V_k at $A \in \mathcal{K}^n$ by means of the values of V_k for projections of A on linear subspaces of different dimensions.

As with the Crofton formulae, we can restrict our considerations to linear subspaces of dimension $n - 1$.

Let $\mathcal{G}^n := \mathcal{G}^n_{n-1}$ be the Grassmannian, that is, the family of linear hyperplanes in $\mathbb{R}^n$. Evidently, $\mathcal{G}^n$ is a subset of $\mathcal{E}^n$:

$$\mathcal{G}^n = \{E \in \mathcal{E}^n \mid \exists v \in S^{n-1} \; E = \phi(v, 0)\};$$

hence in $\mathcal{G}^n$ we have the subspace topology induced by the limit (Definition 2.5.2). We define the measure $\nu : \mathcal{B}(\mathcal{G}^n) \to \mathbb{R}$:
for every $\mathcal{X} \in \mathcal{B}(\mathcal{G}^n)$,

$$\nu(\mathcal{X}) := \sigma\{v \in S^{n-1} \mid \phi(v, 0) \in \mathcal{X}\}. \tag{9.10}$$

Now, with any functional $\Phi : \mathcal{K}^n \to \mathbb{R}$ we associate $\tilde{\Phi} : \mathcal{K}^n \to \mathbb{R}$ defined by the formula

$$\tilde{\Phi}(A) := \int_{\mathcal{G}^n} \Phi(\pi_E(A))d\nu(E). \tag{9.11}$$

9.3.1. THEOREM. *For every* $\Phi : \mathcal{K}^n \to \mathbf{R}$,
(i) *if* Φ *is continuous, then* $\tilde{\Phi}$ *is also continuous;*
(ii) *if* Φ *is a valuation, then* $\tilde{\Phi}$ *is also a valuation;*
(iii) *if* Φ *is invariant, then* $\tilde{\Phi}$ *is also invariant.*

Proof. Since the orthogonal projection π_E on a linear subspace E is a weak contraction, it follows that for every $A \in \mathcal{K}^n$ and $\alpha > 0$,

$$\pi_E((A)_\alpha) \subset (\pi_E(A))_\alpha.$$

Thus in view of (1.4),

$$A = \lim_H A_k \implies \pi_E(A) = \lim_H \pi_E(A_k),$$

and thus condition (i) holds.

By additivity of the integral, to verify (ii) it suffices to prove that for every $A_1, A_2 \in \mathcal{K}^n$ with convex union $A_1 \cup A_2$,

$$\pi_E(A_1 \cup A_2) = \pi_E(A_1) \cup \pi_E(A_2) \tag{9.12}$$

and

$$\pi_E(A_1 \cap A_2) = \pi_E(A_1) \cap \pi_E(A_2). \tag{9.13}$$

Condition (9.12) holds for arbitrary A_1, A_2 (this is a particular case of the formula for the image of the union under any function). Similarly, the inclusion $\subset$ in (9.13) is satisfied by arbitrary A_1, A_2. For the inclusion $\supset$, the convexity of the union is essential (Exercise 2.11).

$\supset$: Obviously, we may assume that

$$A_1 \cap A_2 \neq A_i \text{ for } i = 1, 2. \tag{9.14}$$

Let

$$y \in \pi_E(A_1) \cap \pi_E(A_2).$$

Then there exist x_1, x_2 such that $\pi_E(x_1) = y = \pi_E(x_2)$. If $x_1 = x_2$, then $y \in \pi_E(A_1 \cap A_2)$. If $x_1 \neq x_2$, then

$$\Delta(x_1, x_2) \subset \pi_E^{-1}(y) \cap (A_1 \cup A_2),$$

because $A_1 \cup A_2$ and $\pi_E^{-1}(y)$ are convex (the second set is a line). By (9.14), the set $A_1 \cap A_2$ disconnects $A_1 \cup A_2$; hence

$$\Delta(x_1, x_2) \cap (A_1 \cap A_2) \neq \emptyset.$$

Let $x \in \Delta(x_1, x_2) \cap (A_1 \cap A_2)$; then $x \in \pi_E^{-1}(y)$, whence

$$\pi_E(x) = y \in \pi_E(A_1 \cap A_2).$$

This completes the proof of (9.13).

Condition (iii) is obvious. $\square$

9.3.2. THEOREM. *Let* $n \geq 2$. *For* $k = 0, \ldots, n - 1$, *let*

$$\gamma_{n,k} := \frac{1}{n - k} \cdot \frac{\kappa_{n-k-1}}{\kappa_{n-1}\kappa_{n-k}}.$$

Then for every $A \in \mathcal{K}^n$,

$$V_k(A) = \gamma_{n,k} \int_{\mathcal{G}^n} V_k(\pi_E(A))dv(E).$$

Proof. Let $\Phi := V_k$. Then according to (9.11),

$$\tilde{\Phi}(A) = \int_{\mathcal{G}^n} V_k(\pi_E(A))dv(E). \tag{9.15}$$

By 7.3.3 combined with 9.3.1, the functional $\tilde{\Phi}$ is a continuous invariant valuation, whence in view of the Hadwiger Theorem 8.1.5, there exist $\alpha_0, \ldots, \alpha_n$ such that

$$\tilde{\Phi} = \sum_{i=0}^{n} \alpha_i V_i. \tag{9.16}$$

We shall determine $\alpha_0, \ldots, \alpha_n$, calculating $\tilde{\Phi}(rB^n)$ for arbitrary positive r. By (9.15) and (9.10),

$$\tilde{\Phi}(rB^n) = \int_{S^{n-1}} V_k(\pi_{\phi(v,0)}(rB^n))d\sigma(v).$$

Since the orthogonal projection of rB^n on any hyperplane is a ball of dimension $n - 1$ with the same radius, by the homogeneity of degree k of V_k, from 7.2.13 combined with 7.3.4 it follows that

$$\tilde{\Phi}(rB^n) = n\kappa_n \binom{n-1}{k} \frac{\kappa_{n-1}}{\kappa_{n-k-1}} r^k.$$

On the other hand, by (9.16) and 7.2.13,

$$\tilde{\Phi}(rB^n) = \sum_{i=0}^{n} \alpha_i \binom{n}{i} \frac{\kappa_n}{\kappa_{n-i}} r^i.$$

Comparing coefficients of corresponding powers of the variable r, we obtain

$$\alpha_i = \begin{cases} 0 & \text{if } i \neq k, \\ (n - k)\dfrac{\kappa_{n-1}\kappa_{n-k}}{\kappa_{n-k-1}} & \text{if } i = k. \end{cases}$$

To complete the proof, it now remains to insert $\alpha_0, \ldots, \alpha_n$ into (9.16). $\square$

In particular, for $n = 3$, we obtain the classical Cauchy formulae. They can be written in terms of the functionals M and F for $\mathcal{K}^3$ defined by (7.22) and the functionals f and l for $\mathcal{K}^2$, where $f(A)$ is the area of A and $l(A)$ is the perimeter of A (compare with Exercise 9.5).

9.3.3. THEOREM. *Let $A \in \mathcal{K}^3$. Then*

$$M(A) = \frac{1}{2\pi} \int_{\mathcal{G}^3} l(\pi_E(A))dv(E),$$

$$F(A) = \frac{1}{\pi} \int_{\mathcal{G}^3} f(\pi_E(A))dv(E).$$

Part II

10

Curvature and Surface Area Measures

Every set $A \in \mathcal{K}^n$ determines two finite sequences of measures: curvature measures $\Phi_i(A, \cdot) : \mathcal{B}(\mathrm{R}^n) \to \mathrm{R}$ for $i = 0, \ldots, n$ and surface area measures $\Psi_i(A, \cdot) : \mathcal{B}(S^{n-1}) \to \mathrm{R}$ for $i = 0, \ldots, n$.

For each i, both Φ_i and Ψ_i are "localizations" of intrinsic volume V_i (see Theorem 10.1.7 (b) and 10.2.6):

$$\Phi_i(A, \mathrm{R}^n) = V_i(A) = \Psi_i(A, S^{n-1}).$$

The notion of surface area measures was introduced independently by A.D. Alexandrov in 1937 and by W. Fenchel and B. Jessen in 1938 ([1] and [19]). The notion of curvature measures was introduced by H. Federer in 1959 for a larger class of sets than $\mathcal{K}^n$ ([18]); this class will be the subject of Section 11.1. Twenty years later, R. Schneider in [62] developed the theory of curvature measures for convex compact sets. Also, his survey in [63] deserves particular attention.

10.1 Curvature measures

For any $A \in \mathcal{K}^n$ and $\varepsilon > 0$, the set

$$\mathrm{col}_\varepsilon A := (A)_\varepsilon \setminus A$$

will be called the *ε-collar of A*,

For every $\varepsilon > 0$ we define the function $U_\varepsilon : \mathcal{K}^n \times \mathcal{B}(\mathrm{R}^n) \to \mathrm{R}$:

$$U_\varepsilon(A, X) := \lambda_n((\mathrm{col}_\varepsilon A) \cap \xi_A^{-1}(X \cap A)). \tag{10.1}$$

Thus, $U_\varepsilon(A, X)$ is the volume (that is, n-dimensional Lebesgue measure) of the intersection of the ε-collar of A and the inverse image of $X \cap A$ by the metric projection on A (compare with 3.3.2). Examples are shown in Figure 10.1.[1]

Evidently,

$$U_\varepsilon(A, X) = U_\varepsilon(A, X \cap A) \qquad (10.2)$$

and

$$U_\varepsilon(A, \mathbf{R}^n) = V_n((A)_\varepsilon) - V_n(A). \qquad (10.3)$$

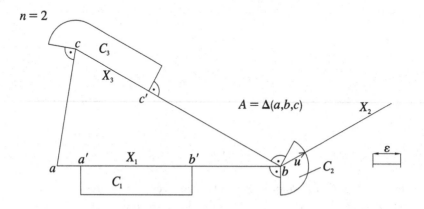

C_i-intersection of $\operatorname{col}_\varepsilon A$ and $\xi_A^1 (X_i \cap A)$ for $i = 1,2,3$
$X_1 = \Delta(a',b') \subset \operatorname{relint} \Delta(a,b)$
$X_2 = b + \operatorname{pos} u$
$X_3 = \Delta(c,c'), \quad c' \in \operatorname{relint} \Delta(b,c)$

Figure 10.1.

Let us note the following.

10.1.1. PROPOSITION. *For every $A \in \mathcal{K}^n$ and $\varepsilon > 0$, the function $U_\varepsilon(A, \cdot)$: $\mathcal{B}(\mathbf{R}^n) \to \mathbf{R}$ is a finite measure on $\mathbf{R}^n$.*

Proof. Obviously, this function is nonnegative and bounded (by (10.2)). Directly from definition (10.1) it follows that for every sequence $(X_i)_{i \in \mathbf{N}}$ of pairwise disjoint Borel sets in $\mathbf{R}^n$,

$$U_\varepsilon\left(A, \bigcup_i X_i\right) = \sum_i U_\varepsilon(A, X_i). \qquad \square$$

We present without proof the following theorem on weak convergence of measures (see (3.3) in [62]):

---

[1] In [62], $U_\varepsilon(A, X)$ is the volume of intersection of $(A)_\varepsilon$ and $\xi_A^{-1}(X \cap A)$.

10.1.2. PROPOSITION. *If $A = \lim_H A_k$, then for every $\varepsilon > 0$,*

$$U_\varepsilon(A_k, \cdot) \xrightarrow{w} U_\varepsilon(A, \cdot).$$

As we did for intrinsic volumes (Definition 7.2.4), we first define curvature measures for the class $\mathcal{P}^n$:

10.1.3. DEFINITION. For every $P \in \mathcal{P}^n$ and $X \in \mathcal{B}(\mathbf{R}^n)$,

$$\Phi_i(P, X) := \begin{cases} \sum_{F \in \mathcal{F}^i(P)} \mathcal{H}^i(F \cap X)\gamma(P, F) & \text{for } i = 0, \ldots, n-1, \\ \lambda_n(P \cap X) & \text{for } i = n. \end{cases}$$

(compare with Definition 7.2.1.)

Evidently,

$$\Phi_i(P, \mathbf{R}^n) = V_i(P). \tag{10.4}$$

By (10.3) and (10.4), the following theorem is a generalization of the Steiner theorem for polytopes, 7.2.9. Its proof is analogous to that of 7.2.9.

10.1.4. THEOREM. *For every $P \in \mathcal{P}^n$, $X \in \mathcal{B}(\mathbf{R}^n)$ and $\varepsilon > 0$,*

$$U_\varepsilon(P, X) = \sum_{i=0}^{n-1} \varepsilon^{n-i}\kappa_{n-i}\Phi_i(P, X). \tag{10.5}$$

According to Definition 10.1.3, the functions $\Phi_0, \ldots, \Phi_n$ are defined on the Cartesian product $\mathcal{P}^n \times \mathcal{B}(\mathbf{R}^n)$. We shall now extend them over $\mathcal{K}^n \times \mathcal{B}(\mathbf{R}^n)$. To this end, let us substitute $\varepsilon = 1, \ldots, n$ in formula (10.5). It is then evident that $(U_1(P, X), \ldots, U_n(P, X))$ is the image of the vector $(\Phi_0(P, X), \ldots, \Phi_{n-1}(P, X))$ under the linear transformation with matrix $M(n, \ldots, 1)\Pi_{j=1}^n \kappa_j$, which is nonsingular as the product of the Vandermonde matrix $M(n, \ldots, 1)$ and a number different from 0. Hence, for every $i \in \{0, \ldots, n-1\}$ there exists (a unique) sequence $(\alpha_{i,1}, \ldots, \alpha_{i,n})$, independent of P and X, such that

$$\Phi_i(P, X) = \sum_{j=1}^n \alpha_{i,j} U_j(P, X).$$

Now let for every $A \in \mathcal{K}^n$ and $X \in \mathcal{B}(\mathbf{R}^n)$,

$$\Phi_i(P, X) := \begin{cases} \sum_{j=1}^n \alpha_{i,j} U_j(P, X) & \text{for } i = 0, \ldots, n-1, \\ \lambda_n(A \cap X) & \text{for } i = n. \end{cases} \tag{10.6}$$

As a direct consequence of 10.1.2, we obtain the following.

10.1.5. PROPOSITION. *If $A = \lim_H A_k$, then for every $i \in \{0, \ldots, n\}$,*

$$\Phi_i(A_k, \cdot) \xrightarrow{w} \Phi_i(A, \cdot).$$

Let us prove

10.1.6. THEOREM. *Let $A \in \mathcal{K}^n$.*
(i) *For $i \in \{0, \ldots, n\}$ the function $\Phi_i(A, \cdot) : \mathcal{B}(\mathbb{R}^n) \to \mathbb{R}$ is a measure.*
(ii) *For every $\varepsilon > 0$ and $X \in \mathcal{B}(\mathbb{R}^n)$,*

$$U_\varepsilon(A, X) = \sum_{i=0}^{n-1} \varepsilon^{n-i} \kappa_{n-i} \Phi_i(A, X).$$

Proof. (i) follows from 10.1.1 combined with (10.6).

(ii): We approximate A by a sequence of polytopes and apply, in turn, 10.1.2, 10.1.4, and 10.1.5.   □

The function $\Phi_i(A, \cdot)$ is the *i-dimensional curvature measure* (or *curvature measure of order i*) of A.

10.1.7. PROPOSITION. (a) *For every $A \in \mathcal{K}^n$, $X \in \mathcal{B}(\mathbb{R}^n)$, and $i \in \{0, \ldots, n-1\}$,*

$$\Phi_i(A, X) = \Phi_i(A, X \cap \mathrm{bd}A).$$

(b) *For every $A \in \mathcal{K}^n$ and i,*

$$\Phi_i(A, \mathbb{R}^n) = V_i(A).$$

Proof. Condition (a) follows from (10.6) combined with (10.2); condition (b) follows from (10.6) combined with (7.10) (the Steiner formula for $\mathcal{K}^n$).   □

As we shall see below (Theorem 10.1.11), curvature measures of orders 0 and $n-1$ have a simple geometric interpretation ([62], (3.20) and (3.21)).

Let $A(u)$ be the support set of A in the direction of a vector u (compare Section 2.2):

$$A(u) := A \cap H(A, u).$$

10.1.8. DEFINITION. Let $A \in \mathcal{K}^n$ and $X \in \mathcal{B}(\mathbb{R}^n)$.

The *spherical image of X with respect to A*, in symbols $\sigma(A, X)$, is defined by the formula

$$\sigma(A, X) := \{u \in S^{n-1} \mid A(u) \cap X \neq \emptyset\}.$$

It is easy to check that this definition is consistent with that given by Schneider in [64] (compare with Exercise 10.2). Let us observe (Exercise 10.3) the following simple relationship between the notion of spherical image and that of outer normal angle (see Definition 7.2.1):

10.1.9. PROPOSITION. *If $P \in \mathcal{P}^n$ and F is a proper face of P, then*

$$\sigma(P, \mathrm{relint}F) = \mathrm{nor}(P, F).$$

In particular, for $\dim F = 0$ the function γ (see Definition 7.2.1) can be extended in an obvious way over $\mathcal{K}^n \times \mathcal{B}(\mathbb{R}^n)$ (compare Exercise 10.4):

10.1.10. DEFINITION. For $A \in \mathcal{K}^n$ and $X \in \mathcal{B}(\mathbb{R}^n)$, let

$$\gamma_0(A, X) := \frac{\sigma_{n-1}(\sigma(A, X))}{\omega_n}.$$

10.1.11. THEOREM. *For every $A \in \mathcal{K}^n$ and $X \in \mathcal{B}(\mathbb{R}^n)$*,

$$\Phi_0(A, X) = \gamma_0(A, X), \tag{10.7}$$

$$\Phi_{n-1}(A, X) = \begin{cases} \frac{1}{2}\mathcal{H}^{n-1}(\mathrm{bd}A \cap X) & \text{if int } A \neq \emptyset \\ \mathcal{H}^{n-1}(A \cap X) & \text{if int } A = \emptyset. \end{cases} \tag{10.8}$$

The following theorem describes basic properties of curvature measures.

10.1.12. THEOREM. *For every $i \in \{0, \ldots, n\}$*,
(i) $\Phi_i : \mathcal{K}^n \times \mathcal{B}(\mathbb{R}^n) \rightarrow \mathbb{R}$ *is invariant under isometries: for any isometry* $f : \mathbb{R}^n \rightarrow \mathbb{R}^n$,

$$\Phi_i(f(A), f(X)) = \Phi_i(A, X);$$

(ii) Φ_i *is homogeneous of degree i: for every $t > 0$*,

$$\Phi_i(tA, tX) = t^i \Phi_i(A, X);$$

(iii) *for every $X \in \mathcal{B}(\mathbb{R}^n)$, the function $\Phi_i(\cdot, X) : \mathcal{K}^n \rightarrow \mathbb{R}$ is a valuation.*

The proof of (i) and (ii) is based only on the definition of curvature measures (Exercise 10.5). For the proof of (iii) the reader is referred to [62].

The curvature measures are *defined locally*. It means that they have the property described by the following theorem.

10.1.13. THEOREM. *Let $A_1, A_2 \in \mathcal{K}^n$ and let X be an open subset of $\mathbb{R}^n$. If*

$$X \cap A_1 = X \cap A_2,$$

then

$$\forall X' \in \mathcal{B}(\mathbb{R}^n) \ \ X' \subset X \implies \Phi_i(A_1, X') = \Phi_i(A_2, X') \ \text{ for } i = 0, \ldots, n.$$

Let us observe that in Theorem 10.1.13 the assumption that X is open is essential:

10.1.14. EXAMPLE. Let a be a common vertex of polytopes P_1 and P_2 and let $P_1 \setminus \{a\} \subset \mathrm{int}P_2$. Then by (10.7),

$$\Phi_0(P_1, \{a\}) = \gamma_0(P_1, \{a\}) \neq \gamma_0(P_2, \{a\}) = \Phi_0(P_2, \{a\}).$$

The Hadwiger Theorem 8.1.5 characterizes linear combinations of basic functionals as the continuous invariant valuations. In Theorem 10.1.15, an analogue of 8.1.5 for curvature measures, valuations with values in $\mathbb{R}$ are replaced by valuations whose values are Borel measures.

A function ϕ from $\mathcal{K}^n$ in the set of Borel measures on $\mathbf{R}^n$ is

- a *valuation* if and only if for every $X \in \mathcal{B}(\mathbf{R}^n)$ the function $A \mapsto \phi(A)(X)$ is a valuation,
- *weakly continuous* if and only if

$$A = \lim_H A_k \Longrightarrow \phi(A_k) \overset{w}{\to} \phi(A),$$

- *invariant* (*under isometries*) if and only if

$$\phi(f(A))(f(X)) = \phi(A)(X)$$

for every isometry $f : \mathbf{R}^n \to \mathbf{R}^n$,
- *defined locally* if and only if for every $A \in \mathcal{K}^n$ the measure $\phi(A)$ is defined locally.

10.1.15. THEOREM. *For any function ϕ from $\mathcal{K}^n$ into the set of Borel measures on $\mathbf{R}^n$, the following conditions are equivalent:*
(i) *ϕ is a weakly continuous, locally defined, invariant valuation;*
(ii) *there exist $\alpha_0, \ldots, \alpha_n \geq 0$ such that for every $A \in \mathcal{K}^n$ and $X \in \mathcal{B}(\mathbf{R}^n)$,*

$$\phi(A)(X) = \sum_{i=0}^{n} \alpha_i \Phi_i(A, X).$$

It is easy to prove that the set of weakly continuous, locally defined, invariant valuations is closed under addition and multiplication by nonnegative scalars (Exercise 10.6). Thus the implication (ii) $\Longrightarrow$ (i) in Theorem 10.1.15 follows from 10.1.5, 10.1.12, and 10.1.13. We omit the (difficult) proof of the converse implication.

Also, the Crofton formulae (Theorem 9.2.6) have their counterpart for curvature measures. As in Section 9.2, we restrict our consideration to sections by hyperplanes. We keep in mind the notation introduced there.

10.1.16. THEOREM. *Let $n \geq 2$. For $i \in \{1, \ldots, n\}$, let*

$$\beta_{n,i} := \frac{2}{i} \cdot \frac{\kappa_{i-1}}{\kappa_i \kappa_{n-1}}.$$

Then for every $A \in \mathcal{K}^n$ and $X \in \mathcal{B}(\mathbf{R}^n)$,

$$\Phi_i(A, X) = \beta_{n,i} \int_{\mathcal{E}_A} \Phi_{i-1}(A \cap E, X) d\mu(E).$$

Applying (10.8), it is easy to prove that $(n-1)$-dimensional curvature measure $\Phi_{n-1}(A, \cdot)$ determines uniquely the set A (Exercise 10.7).

The natural question arises whether the isometry group is the maximal group of transformations of $\mathbf{R}^n$ preserving curvature measures.

10.1.17. PROBLEM. Prove or disprove the following: if f is a homeomorphism of R^n satisfying the condition

$$\forall X \in \mathcal{B}(R^n) \ \forall i \in \{0, \dots, n-1\} \ \Phi_i((f(A), f(X)) = \Phi_i(A, X) \qquad (10.9)$$

for some $A \in \mathcal{K}_0^n$, then $f|\text{bd}A$ is an isometry of $\text{bd}A$ onto $\text{bd}f(A)$.

The following Schneider theorem ([62], (9.1)) gives a partial solution to Problem 10.1.17 (see Exercise 10.8).

10.1.18. THEOREM. *Let $A_1, A_2 \in \mathcal{K}^n$, $0 \in \text{int}(A_1 \cap A_2)$, and let $f : \text{bd}A_1 \to \text{bd}A_2$ be the central projection from 0.*
If there exists an $i \in \{1, \dots, n-1\}$ such that

$$\Phi_i(A_1, X) = \Phi_i(A_2, f(X))$$

for every Borel set $X \subset \text{bd}A_1$, then $A_1 = A_2$.

The proof of Theorem 10.1.18 requires a more general version of Theorem 10.1.16, namely the version concerning sections by affine subspaces of arbitrary dimension $k < n$. We confine ourselves to the proof of a weaker theorem, 10.1.18'.

Let us first notice that in Theorem 10.1.16 the integral over the set $\mathcal{E}_A$ of hyperplanes intersecting A can be replaced by the integral over the whole set $\mathcal{E}^n$ if we allow $A \cap E$ to be empty and define

$$\Phi_i(\emptyset, X) := 0$$

for every i.

10.1.18'. THEOREM. *Let $A_1, A_2 \in \mathcal{K}^n$, $0 \in \text{int}(A_1 \cap A_2)$, and let $f : \text{bd}A_1 \to \text{bd}A_2$ be the central projection from 0.*
If for every Borel set $X \subset \text{bd}A_1$,

$$\Phi_1(A_1, X) = \Phi_1(A_2, f(X)),$$

then $A_1 = A_2$.

Proof. Suppose $A_1 \neq A_2$. Then

$$\text{bd}A_1 \setminus A_2 \neq \emptyset \ \text{ or } \ \text{bd}A_2 \setminus A_1 \neq \emptyset,$$

because otherwise $\text{bd}A_1 \subset A_2$ and $\text{bd}A_2 \subset A_1$, whence $A_1 = A_2$, a contradiction. (Of course, we make use of the assumption that A_1 and A_2 are convex.)

Thus, we may assume that

$$X := \text{bd}A_1 \setminus A_2 \neq \emptyset.$$

Of course, $X \in \mathcal{B}(R^n)$. We shall first prove that for every $E \in \mathcal{E}^n$,

$$\Phi_0(A_1 \cap E, X) \geq \Phi_0(A_2 \cap E, f(X)). \qquad (10.10)$$

Notice that $f(X) \subset \operatorname{int} A_1$; hence

$$A_1 \cap E = \emptyset \Longrightarrow f(X) \cap E = \emptyset \Longrightarrow \Phi_0(A_2 \cap E, f(X)) = 0$$

(compare with 10.1.7 (a)). Therefore, if $A_1 \cap E = \emptyset$, then (10.10) is satisfied.

Let $A_1 \cap E \neq \emptyset \neq A_2 \cap E$. Then for every $(n-2)$-dimensional hyperplane $H(A_2 \cap E, u)$ in E supporting $A_2 \cap E$ at some point of $f(X)$, there exists the parallel $(n-2)$-dimensional hyperplane $H(A_1 \cap E, u)$ in E supporting $A_1 \cap E$ at some point of X. Hence in this case condition (10.10) is satisfied as well.

Finally, if

$$A_1 \cap E \neq \emptyset = A_2 \cap E, \tag{10.11}$$

then

$$\Phi_0(A_1 \cap E, X) \neq 0 = \Phi_0(A_2 \cap E, f(X)),$$

whence in (10.10) the inequality is sharp. Since the set of hyperplanes satisfying (10.11) is of positive measure μ, in view of Theorem 10.1.16 it follows that

$$\Phi_1(A_1, X) > \Phi_1(A_2, f(X)),$$

contrary to the assumption. $\qquad\qquad\qquad\qquad\qquad\qquad\qquad\qquad\qquad$ $\square$

Let us complete this section with some examples.

10.1.19. EXAMPLE. Let $n = 2$. Let $A := \Delta(a, b, c)$ and X_1, X_2, X_3 be as in Figure 10.1. We calculate $\Phi_i(A, X_j)$ for $i = 0, 1$ and $j = 1, 2, 3$.

By 10.1.3 combined with 7.2.1,

$$\Phi_i(A, X_j) = \sum\nolimits_{F \in \mathcal{F}^i(A)} \mathcal{H}^i(F \cap X_j)(\omega_{2-i})^{-1} \sigma_{1-i}(\operatorname{nor}(A, F)),$$

where $\operatorname{nor}(A, F) = \{u \in S^{n-1} \mid A(u) = F\}$. Thus it is easy to see that

$$\Phi_0(A, X_1) = 0,$$

$$\Phi_0(A, X_2) = \frac{1}{2\pi}(\pi - \angle(a, b, c)),$$

$$\Phi_0(A, X_3) = \frac{1}{2\pi}(\pi - \angle(a, c, b)),$$

and

$$\Phi_1(A, X_1) = \frac{1}{2}\|a' - b'\|, \quad \Phi_1(A, X_2) = 0, \quad \Phi_1(A, X_3) = \frac{1}{2}\|c - c'\|.$$

10.1.20. EXAMPLE. Let $n = 3$. Consider the cylinder

$$A = \{(x_1, x_2, x_3) \in \mathbb{R}^3 \mid x_1^2 + x_2^2 \leq 1, \ 0 \leq x_3 \leq 1\}$$

and let $X_1 = A \cap \operatorname{aff}(e_1, e_2)$, $X_2 = \operatorname{relbd} X_1$, and $X_3 = \Delta((1, 0, 0), (1, 0, 1))$.

To calculate the curvature measures $\Phi_i(X_j)$, for $i = 0, 1, 2$ and $j = 1, 2, 3$, we apply 10.1.6(ii). Take $\varepsilon \in (0; \frac{1}{2})$. The counterimage of X_j under the metric projection onto A is the half-space $H = \{(x_1, x_2, x_3) \mid x_3 \leq 0\}$ for $j = 1$, the set $H \setminus \{(x_1, x_2, x_3) \mid x_1^2 + x_2^2 \leq 1, \ -\varepsilon \leq x_3 \leq 0\}$ for $j = 2$, and the half-plane $\{(x_1, 0, x_3) \mid x_1 \geq 1\}$ for $j = 3$.

To calculate the volumes of intersections of $\mathrm{col}_\varepsilon(A)$ with those three counterimages, we use the Guldin formula for volume of the set obtained by rotating a plane figure D about the axis disjoint from D (here the axis $\mathrm{lin}(e_3)$).

For $j = 1$, we obtain

$$U_\varepsilon(A, X_1) = \pi\varepsilon + \frac{\pi^2\varepsilon^2}{2} + \frac{2}{3}\pi\varepsilon^3;$$

hence by 10.1.6(ii),

$$\pi\varepsilon + \frac{\pi^2\varepsilon^2}{2} + \frac{2}{3}\pi\varepsilon^3 = \varepsilon^3\frac{4}{3}\pi\,\Phi_0(A, X_1) + \varepsilon^2\pi\,\Phi_1(A, X_1) + 2\varepsilon\Phi_2(A, X_1).$$

Comparing the coefficients of corresponding powers of ε on both sides of the equation above, we obtain $\Phi_0(A, X_1) = \frac{1}{2}$ and $\Phi_i(A, X_1) = \frac{\pi}{2}$ for $i = 1, 2$.

For $j = 2$,

$$\xi_A^{-1}(X_2) = H \setminus \{(x_1, x_2, x_3) \mid (x_1, x_2, 0) \in \mathrm{relint}X_1\},$$

whence

$$U_\varepsilon(A, X_2) = \frac{1}{2}\pi^2\varepsilon^2 + \frac{2}{3}\pi\varepsilon^3.$$

Thus, comparing again the coefficients, we obtain

$$\Phi_0(A, X_2) = \frac{1}{2}, \quad \Phi_1(A, X_2) = \frac{\pi}{2}, \quad \Phi_2(A, X_2) = 0.$$

For $j = 3$, since the left side of the formula in 10.6 (ii) is 0, we obtain $\Phi_i(A, X_3) = 0$ for $i = 0, 1, 2$.

We leave to the reader the details of the above calculations (see Exercise 10.11).

10.2 Surface area measures

Unlike the curvature measures, which are Borel measures on $\mathbf{R}^n$, the surface area measures are Borel measures on S^{n-1}.

For every $A \in \mathcal{K}^n$, the function $u_A : \mathbf{R}^n \setminus A \to S^{n-1}$ is defined by

$$u_A(x) := \frac{x - \xi_A(x)}{\varrho(x, A)}$$

(compare [64], p. 9).

From 1.1.2 and 3.3.4 it follows that

10.2.1. *For every $A \in \mathcal{K}^n$ the function u_A is continuous.*

By 10.2.1, inverse images (with respect to u_A) of Borel sets are Borel sets:

$$Y \in \mathcal{B}(S^{n-1}) \implies u_A^{-1}(Y) \in \mathcal{B}(\mathbb{R}^n).$$

Thus for every $\varepsilon > 0$ the function $V_\varepsilon : \mathcal{K}^n \times \mathcal{B}(S^{n-1}) \to \mathbb{R}$ may be defined by the formula

$$V_\varepsilon(A, Y) := \lambda_n((\mathrm{col}_\varepsilon(A)) \cap u_A^{-1}(Y)) \qquad (10.12)$$

(Figure 10.2).

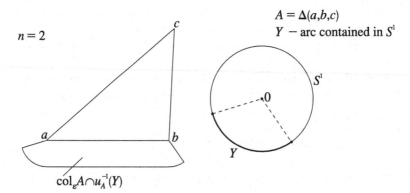

Figure 10.2.

Evidently, directly from definition (10.12) we obtain an analogue of (10.3):

$$V_\varepsilon(A, S^{n-1}) = V_n(A_\varepsilon) - V_n(A). \qquad (10.13)$$

The following statement is a counterpart of Theorem 10.1.1 concerning $U_\varepsilon(A, \cdot)$.

10.2.2. *For every $A \in \mathcal{K}^n$ and $\varepsilon > 0$, the function $V_\varepsilon(A, \cdot)$ is a finite measure on S^{n-1}.*

As with the curvature measures (Definition 10.1.3), the surface area measures are defined first for convex polytopes:

10.2.3. DEFINITION. For every $P \in \mathcal{P}^n$, $Y \in \mathcal{B}(S^{n-1})$, and $i \in \{0, \ldots, n-1\}$

$$\Psi_i(P, Y) := \sum_{F \in \mathcal{F}^i(P)} \frac{\sigma_{n-1-i}(\sigma(P, F) \cap Y)}{\omega_{n-i}} \mathcal{H}^i(F).$$

In view of (10.13), the following theorem is a generalization of the Steiner theorem for polytopes (Theorem 10.1.4).

10.2.4. THEOREM. *For every* $P \in \mathcal{P}^n$, $Y \in \mathcal{B}(S^{n-1})$, *and* $\varepsilon > 0$,

$$V_\varepsilon(P, Y) = \sum_{i=0}^{n-1} \varepsilon^{n-i} \kappa_{n-i} \Psi_i(P, Y).$$

For every $i \in \{0, \ldots, n-1\}$, the function $\Psi_i : \mathcal{P}^n \times \mathcal{B}(S^{n-1}) \to \mathbb{R}$ can be extended over $\mathcal{K}^n \times \mathcal{B}(S^{n-1})$ (compare Section 10.1). As before, we use the same symbol for the extension as for the original function.

The following statement is a counterpart of 10.1.6.

10.2.5. THEOREM. *For every* $A \in \mathcal{K}^n$ *and* $i \in \{0, \ldots, n-1\}$, *the function* $\Psi_i(A, \cdot)$ *is a finite measure on* S^{n-1} *and for every* $\varepsilon > 0$,

$$V_\varepsilon(A, \cdot)) = \sum_{i=0}^{n-1} \varepsilon^{n-i} \kappa_{n-i} \Psi_i(A, \cdot).$$

The function $\Psi_i(A, \cdot)$ is called the *i-dimensional surface area measure of A.*

The following statement is a counterpart of 10.1.7 (b).

10.2.6. PROPOSITION. *For every* $A \in \mathcal{K}^n$,

$$\Psi_i(A, S^{n-1}) = V_i(A).$$

10.2.7. DEFINITION. Let $A \in \mathcal{K}^n$ and $Y \in \mathcal{B}(S^{n-1})$. The *reverse spherical image of Y with respect to A* is the set $\tau(A, Y)$ defined by

$$\tau(A, Y) := \bigcup_{u \in Y} A(u).$$

The surface area measures are *defined locally* but in a more general sense than the curvature measures (compare 10.1.13):

10.2.8. THEOREM. *Let* $A_1, A_2 \in \mathcal{K}^n$ *and* $Y \in \mathcal{B}(S^{n-1})$. *If*

$$\tau(A_1, Y) = \tau(A_2, Y),$$

then for every i,

$$\Psi_i(A_1, Y) = \Psi_i(A_2, Y).$$

Further, let us note an analogue of 10.1.12 (ii):

10.2.9. PROPOSITION. *For every* $Y \in \mathcal{B}(S^{n-1})$, *the function* $\Psi_i(\cdot, Y)$ *is homogeneous of degree i:*

$$\forall t > 0 \quad \Psi_i(tA, Y) = t^i \Psi_i(A, Y).$$

A classical theorem of Alexandrov, Fenchel, and Jessen (1937, 1938) concerns determination (up to a translation) of a convex body by the surface area measures ([63]). Like the previous statements in this section, we mention it without proof.

10.2.10. THEOREM. *Let* $j \in \{1, \ldots, n-1\}$. *Let* $A_1, A_2 \in \mathcal{K}^n$, *and let* $\dim A_i > j$ *for* $i = 1, 2$. *If*

$$\Psi_j(A_1, \cdot) = \Psi_j(A_2, \cdot),$$

then

$$\exists v \in \mathbf{R}^n \ \ A_2 = A_1 + v.$$

(Compare Exercise 10.7 and Theorem 10.1.18 on curvature measures.)

10.3 Curvature and surface area measures for smooth, strictly convex bodies

Let $A \in \mathcal{K}_0^n$. A point $a \in \mathrm{bd}A$ is a *singular point of* A if A has more than one support hyperplane at a, or equivalently, $\dim \sigma(A, \{a\}) > 0$. Otherwise, a is a *regular point of* A. Thus, for any regular point a there is a unique support hyperplane of A at a; that is, the spherical image $\sigma(A, \{a\})$ is a singleton.

Similarly, the direction of $u \in S^{n-1}$ is a *singular direction for* A if the support set $A(u)$ has dimension at least one. Otherwise, it is a *regular direction for* A. Thus if the direction of u is regular, then $A(u)$ is a singleton.

Points and directions can be classified more precisely by means of the dimension of the spherical image or the dimension of the support set, respectively (see [64]). From the point of view of those classifications, the family $\mathcal{P}_0^n$ has "extreme" properties: for every n-dimensional convex polytope P in $\mathbf{R}^n$ and every $k \in \{0, \ldots, n-1\}$,

- there exist points with k-dimensional spherical image; they are the points of the $(n-1-k)$-dimensional faces of P,
- there exist vectors with k-dimensional support sets; they are the outer normal vectors of the k-dimensional faces of P.

We shall now consider the "opposite extreme": the class of convex bodies with all points and all directions regular.

10.3.1. PROPOSITION. (i) *If all the points of a convex body A are regular, then the formula*

$$\sigma_A(x) := \sigma(A, \{x\})$$

defines a function $\sigma_A : \mathrm{bd}A \to S^{n-1}$.

(ii) *If all directions are regular for A, then the formula*

$$\tau_A(u) := \tau(A, \{u\})$$

defines a function $\tau_A : S^{n-1} \to \mathrm{bd}A$.[2]

---

[2] Compare 10.2.7.

In case (i) we say that A is *smooth*; the function σ_A is then referred to as the *spherical map of A*; in case (ii) we say that A is *strictly convex*; the function τ_A is then referred to as the *reverse spherical map*.

Let us note the following.

10.3.2. PROPOSITION. *If A is a smooth, strictly convex body, then σ_A is a homeomorphism and*

$$\sigma_A^{-1} = \tau_A$$

(Exercise 10.9).

For smooth convex bodies as well as for strictly convex bodies the curvature measures are closely related to the surface area measures (Theorem 4.2.4 in [64]):

10.3.3. THEOREM. *Let $A \in \mathcal{K}_0^n$, $i \in \{0, \ldots, n-1\}$. If A is smooth, then for every $Y \in \mathcal{B}(S^{n-1})$,*

$$\Psi_i(A, Y) = \Phi_i(A, \sigma_A^{-1}(Y));$$

if A is strictly convex, then for every $X \in \mathcal{B}(\mathbb{R}^n)$,

$$\Phi_i(A, X) = \Psi_i(A, \tau_A^{-1}(X)).$$

10.3.4. COROLLARY. *If A is a smooth, strictly convex body in $\mathbb{R}^n$, then for $i \in \{0, \ldots, n-1\}$,*

$$\Phi_i(A, X) = \Psi_i(A, \sigma_A(X \cap \operatorname{bd} A)).$$

Smooth convex bodies are a subject of interest of classical differential geometry. Hence let us express their curvature measures in terms of differential geometry.

Let A be a smooth convex body in $\mathbb{R}^n$ and let $a \in \operatorname{bd} A$. There exists a neighborhood S of a in $\operatorname{bd} A$ that is a smooth sheet, which means that there is a domain U in $\mathbb{R}^{n-1}$ and a homeomorphism $p : U \to S$ of the class C^1 with linearly independent partial derivatives $p_i := \frac{\partial p}{\partial u_i}$ for $i = 1, \ldots, n-1$. Let

$$p_n := \frac{\times (p_1, \ldots, p_{n-1})}{\| \times (p_1, \ldots, p_{n-1})\|},$$

where $\times$ is the vector product in $\mathbb{R}^n$ (see [50]), and let $u_0 := p^{-1}(a)$. Then of course, $p_n(u_0)$ is a unit normal vector of S at a. If the choice of the parametric representation p is suitable, then this vector is the inner normal vector of A, i.e., $-p_n$ is the outer normal vector.

Assume now that p is of the class C^2.

Let L be the normal line at a:

$$L := a + \operatorname{lin}(p_n).$$

For every plane N containing L, the intersection $N \cap S$ is an arc (if S is small enough); it is the so-called *normal section*. This arc is the image $p(L)$ of some arc L contained in U with parametrization $u : (\alpha; \beta) \to U$ of the class C^1; consequently, $S \cap N = \{pu(s) \mid s \in (\alpha; \beta)\}$ and $u_0 = u(s_0)$ for some $s_0 \in (\alpha; \beta)$.

Let $r := pu$. We may assume that r is normalized, i.e., its derivative r' has norm equal to 1. Let

$$r_1 := r'.$$

Then

$$r_1(s) = \sum_{i=1}^{n-1} p_i(u(s))u_i'(s), \qquad (10.14)$$

whence $p_n \perp r_1$; thus $(a; r_1(s_0), p_n(u_0))$ is an (orthogonal) affine basis of the plane N. In the plane N oriented by the choice of this basis, the oriented normal vector $\underline{r_2}$ coincides with p_n; hence by the Frenet formulae for an oriented curve, the oriented curvature $\underline{\kappa}$ of the normal section $S \cap N$ has the form

$$\underline{\kappa} = (r_1)' \circ p_n. \qquad (10.15)$$

Let

$$p_{ij} := \frac{\partial p_i}{\partial u_j}$$

and

$$b_{ij} := p_{ij} \circ p_n.$$

From (10.14) and (10.15) it follows that

$$\underline{\kappa} = \sum_{i,j=1}^{n-1} b_{ij} u_i' u_j'. \qquad (10.16)$$

Hence we have expressed the oriented curvature of the normal section $S \cap N$ as the value of *the second fundamental form* $\sum_{ij} b_{ij} x_i x_j$ for $x_i := u_i'$.

Eigenvalues of the matrix $(b_{ij}(u_0))_{i,j=1,\ldots,n-1}$ are *principal curvatures of S at* u_0.

In the case $n = 3$, principal curvatures are extremal values of the curvature $\underline{\kappa} =: \underline{\kappa}_N$ treated as a function of N.

Let $\underline{\kappa}_1, \ldots, \underline{\kappa}_{n-1}$ be the principal curvatures. In the situation considered, either $\underline{\kappa}_i \geq 0$ for all i or $\underline{\kappa}_i \leq 0$ for all i; since, by the assumption, p_n is the inner normal unit vector, it follows that they are nonnegative.

The functions $H_j : U \to \mathbb{R}$ for $j = 1, \ldots, n-1$ are defined by

$$H_j := \binom{n-1}{j}^{-1} \sum_{\substack{1 \leq i_1, \ldots, i_j \leq n-1 \\ i_\kappa < i_{\kappa+1}}} \underline{\kappa}_{i_1} \cdots \underline{\kappa}_{i_j}.$$

The first one, H_1, is the *mean curvature*, and the last, H_{n-1}, the *Gaussian curvature* of S:

$$H_1 = \frac{1}{n-1}(\underline{\kappa}_1 + \cdots + \underline{\kappa}_{n-1}),$$

$$H_{n-1} = \underline{\kappa}_1 \cdots \underline{\kappa}_{n-1}.$$

Since p is a homeomorphism, the H_j can be treated as functions of a point $x \in S$ and can be integrated over $X \cap A$ with respect to the $(n-1)$-dimensional Hausdorff measure $\mathcal{H}^{n-1}$. Covering bd A by a finite family of closures of pairwise disjoint open subsets, we can treat each H_j as a function over bd A.

The proof of the following theorem can be found in [64], Section 2.5. The symbol C_+^2 used there denotes the class of convex bodies whose boundaries can be parametrized (locally) by functions of the class C^2, and the Gaussian curvature is positive.

10.3.5. THEOREM. *For every $A \in \mathcal{K}^n$ of class C_+^2 and every $X \in \mathcal{B}(\mathbf{R}^n)$,*

$$U_\varepsilon(A, X) = \frac{1}{n} \sum_{i=0}^{n-1} \varepsilon^{n-i} \binom{n}{i} \int_{X \cap \mathrm{bd}A} H_{n-1-i} \, d\mathcal{H}^{n-1}. \tag{10.17}$$

Comparing (10.17) with the Steiner formula for curvature measures (Theorem 10.1.6 (ii)), we obtain a theorem on curvature measures of smooth convex bodies (Theorem 10.3.6). To formulate this theorem, let us replace Φ_i by the following function C_i, also referred to as the *curvature measure*:

$$C_i := n \binom{n}{i}^{-1} \kappa_{n-i} \Phi_i. \tag{10.18}$$

The function C_i is a localization of the functional W_{n-i} (see (7.7)), in the similar sense similar as Φ_i is a localization of V_i (compare with 10.1.7 (b)):

$$C_i(A, \mathbf{R}^n) = n W_{n-i}(A). \tag{10.19}$$

Unlike Φ_i, the function C_i depends on the dimension of the ambient space.

10.3.6. THEOREM. *For every convex body A of class C_+^2 in $\mathbf{R}^n$, for every $X \in \mathcal{B}(\mathbf{R}^n)$, and $i \in \{0, \ldots, n-1\}$,*

$$C_i(A, X) = \int_{X \cap \mathrm{bd}A} H_{n-1-i} d\mathcal{H}^{n-1}.$$

Theorem 10.3.6 combined with (10.18) and (10.19) yields the following corollary concerning the functionals $W_1, \ldots, W_n$ (and thus concerning also the intrinsic volumes) for convex bodies of class C_+^2 (see [64] (4.2.28)).

10.3.7. COROLLARY. *For every $A \in \mathcal{K}^n$ of the class C_+^2,*

$$W_i(A) = \frac{1}{n} \int_{\mathrm{bd}A} H_{i-1} d\mathcal{H}^{n-1}.$$

Hence in particular, for convex bodies of this class, the functional W_2 is the integral of the mean curvature H_1 (compare footnote to (9.1)).

In [64] the reader can also find analogues of Theorem 10.3.6 and Corollary 10.3.7 for surface area measures.

11

Sets with positive reach. Convexity ring

11.1 Sets with positive reach

As was already mentioned at the beginning of Chapter 10, H. Federer defined curvature measures for a larger class of sets than $\mathcal{K}_0^n$: for the so-called *sets with positive reach*.

11.1.1. DEFINITION. For any nonempty closed subset A of $\mathbf{R}^n$, let

$$D(A) := \{x \in \mathbf{R}^n \mid \exists^1 \, a \in A \, \varrho(x, a) = \varrho(x, A)\};$$
$$\text{reach} A := \sup\{r \geq 0 \mid \{x \mid \varrho(x, A) < r\} \subset D(A)\};$$

$\text{reach} A := \infty$ if this upper bound does not exist.

Thus $\text{reach} A$ is the upper bound of radii of the open generalized balls around A, every point of which has only one nearest point in A.

11.1.2. EXAMPLES. (a) Let A be a broken line in $\mathbf{R}^2$: $A = \Delta(a, b) \cup \Delta(b, c)$ for noncollinear points a, b, c with $\|a - b\| = \|b - c\|$.

Of course, $\text{reach} A = 0$, because every point of the bisectrix of the pair of half-lines $b + \text{pos}(a - b)$, $b + \text{pos}(c - b)$, excluding their common point, has two nearest points in A.

(b) Let A be the sphere in $\mathbf{R}^n$, with center 0 and radius $r > 0$. Every point $x \in \mathbf{R}^n \setminus \{0\}$ has exactly one nearest point in A, while for the point 0 every point of A is the nearest point. Hence $D(A) = \mathbf{R}^n \setminus \{0\}$ and $\text{reach} A = r$.

(c) In view of Theorem 3.3.1, if A is a nonempty closed convex subset of $\mathbf{R}^n$, then $\text{reach} A = \infty$.

11.1.3. DEFINITION. A subset A of $\mathbf{R}^n$ is a *set with positive reach* if $0 <$ reach$A \leq \infty$.

Thus, Theorem 3.3.1 can be reformulated as follows:

11.1.4. THEOREM. *A subset A of $\mathbf{R}^n$ is nonempty, closed, and convex if and only if* reach$A = \infty$.

Let us note that Federer's definition of reach is slightly different; namely, he first defines the local reach, reach(A, a), for $a \in A$:

11.1.5. DEFINITION. Let $A \subset \mathbf{R}^n$ and $a \in A$. Then

$$\text{reach}(A, a) := \sup\{r > 0 \mid \varrho(x, a) < r \Longrightarrow x \in D(A)\}.$$

Further, he defines the reach of A as the lower bound of reach(A, a) for $a \in A$. For compact sets his notion of reach coincides with that defined by 11.1.1 above:

11.1.6. *If A is compact, then*

$$\text{reach}A = \inf\{\text{reach}(A, a) \mid a \in A\}.$$

Proof. For every $a \in A$, the set $\{r > 0 \mid \varrho(x, A) < r \Longrightarrow x \in D(A)\}$ is contained in $\{r > 0 \mid \varrho(x, a) < r \Longrightarrow x \in D(A)\}$, whence reach$A \leq$ reach(A, a) for every $a \in A$. Thus reach$A \leq \inf\{\text{reach}(A, a) \mid a \in A\}$.

Let $r = \inf\{\text{reach}(A, a) \mid a \in A\}$.

Since A is compact, it follows that there exists $a \in A$ with $r = \text{reach}(A, a)$; hence $\{x \mid \varrho(x, a) < r\} \subset D(A)$. This implies

$$\{x \mid \varrho(x, A) < r\} \subset D(A),$$

which means that $r \leq$ reach$A$.   □

11.1.7. THEOREM. ([18] Remark 4.2, p. 432)
(i) *The function $A \ni a \mapsto \text{reach}(A, a)$ is continuous.*
(ii) *for every $a \in$ bd$A$,*

$$0 \leq \text{reach}(\text{bd}A, a) \leq \text{reach}(A, a) \leq \infty.$$

Federer defines the set Tan(A, a) of vectors tangent to $A \subset \mathbf{R}^n$ at $a \in A$:

11.1.8. DEFINITION. ([18], 4.3)

$$u \in \text{Tan}(A, a) :\Longleftrightarrow$$

$$u = 0 \text{ or } \forall \varepsilon > 0 \, \exists x \in A \, \, 0 < \|x - a\| < \varepsilon \text{ and } \left| \frac{x - a}{\|x - a\|} - \frac{u}{\|u\|} \right| < \varepsilon.$$

11.1.9. THEOREM. (compare [18] Remark 4.20, p. 450). *Let A be a set with positive reach in $\mathbf{R}^n$. The following conditions are equivalent:*
(i) *A is a k-dimensional manifold,*
(ii) *for every $a \in A$ the set Tan(A, a) is a k-dimensional linear space.*

The same remark 4.20 in [18] contains some conditions sufficient for a smooth k-dimensional manifold to have positive reach.

Already in 1939, H. Weyl in [67] established an analogue of the Steiner theorem for compact k-dimensional submanifolds of $\mathbf{R}^n$, of class C^2, for $1 \leq k \leq n - 1$.

We shall present here without proof the following Federer theorem (compare with 5.6, p. 455, in [18]):

11.1.10. THEOREM. *For every* $A \subset \mathbf{R}^n$ *with positive reach there exists a unique sequence of Borel measures* $(\Phi_0(A, \cdot), \ldots, \Phi_n(A, \cdot))$ *such that for every* A *and every* $X \in \mathcal{B}(\mathbf{R}^n)$, *if* $0 \leq r < \mathrm{reach}A$, *then*

$$\lambda_n((A)_r \cap \xi_A^{-1}(X)) = \sum_{i=0}^{n} r^{n-i} \kappa_{n-i} \Phi_i(A, X).$$

Evidently, by Theorem 10.1.6 combined with Example 11.1.2 (c), for any $A \in \mathcal{K}^n$ the function $\Phi_i(A, \cdot)$ is the ith curvature measure of A. Hence 11.1.10 is a generalization of 10.1.6 (local version of the Steiner theorem) on the class of sets with positive reach.

11.2 Convexity ring

The class $\mathcal{K}^n \cup \{\emptyset\}$ is closed under intersection (compare with 3.2.1), but of course, it is not closed under union. This class can be extended to the family $\mathcal{U}^n$ defined by the formula

$$A \in \mathcal{U}^n :\Longleftrightarrow \exists A_1, \ldots, A_m \in \mathcal{K}^n \cup \{\emptyset\} \quad A = \bigcup_{i=1}^{m} A_i. \tag{11.1}$$

11.2.1. *The family* $\mathcal{U}^n$ *is a ring of sets (i.e., a ring with respect to* $\cup$ *and* $\cap$) *generated by* $\mathcal{K}^n \cup \{\emptyset\}$.

Proof. Since the power set of any nonempty set, thus of $\mathbf{R}^n$ in particular, is a ring with respect to $\cup$ and $\cap$, it suffices to prove that $\mathcal{U}^n$ is closed under these two operations.

Directly from Definition 11.2.1, it follows that $\mathcal{U}^n$ is closed under $\cup$. It is closed under $\cap$ as well:

Let $A = \bigcup_{i=1}^{m} A_i$ and $B = \bigcup_{j=1}^{k} B_j$ for some $A_1, \ldots, A_m, B_1, \ldots, B_k \in \mathcal{K}^n$. Since $A_i \cap B_j \in \mathcal{K}^n \cup \{0\}$ for every $i \in \{1, \ldots, m\}$, $j \in \{1, \ldots, k\}$, and $A \cap B = \bigcup_{i,j} A_i \cap B_j$, it follows that $A \cap B \in \mathcal{U}^n$. $\qquad\square$

The family $\mathcal{U}^n$ will be called the *convexity ring*.[1]

---

[1] Hadwiger in [30] and Schneider in [64] use the name *Konvexring* and *convex ring*, respectively; it seems, however, that their terminology is confusing.

The study of the convexity ring is motivated by the possibility of generalizing on $\mathcal{U}^n$ some theorems concerning $\mathcal{K}^n$, for instance the Crofton Theorem 9.2.6. Of course, to do this, we have to extend basic functionals over $\mathcal{U}^n$.

We need Lemma 11.2.3, whose proof is due to R. Schneider ([64] p. 174). This lemma concerns some special valuations; following Schneider, we call them fully additive:

11.2.2. DEFINITION. Let $\mathcal{F}$ be a nonempty family of subsets of R^n, closed under intersection, and let $\mathcal{U}(\mathcal{F})$ be the family of finite unions of members of $\mathcal{F}$.

A function $\Phi : \mathcal{F} \to R$ is *fully additive* on $\mathcal{F}$ if for every $A_1, \ldots, A_m \in \mathcal{F}$ with $\bigcup_{i=1}^m A_i \in \mathcal{F}$,

$$\Phi\left(\bigcup_{i=1}^m A_i\right) = \sum_{r=1}^m (-1)^{r-1} \sum_{i_1 < \cdots < i_r} \Phi(A_{i_1} \cap \cdots \cap A_{i_r}). \qquad (11.2)$$

Obviously, every fully additive function is a valuation (that is, it satisfies (11.2) for $m = 2$).

We admit the following notation. For every $m \in N$, let

$$S(m) := \{v \subset \{1, \ldots, m\} \mid v \neq \emptyset\} \quad \text{and} \quad |v| := \text{card}v \quad \text{for } v \in S(m).$$

For any $v = \{i_1, \ldots, i_r\} \in S(m)$ and $A_1, \ldots, A_m \in \mathcal{F}$, let

$$A_v := A_{i_1} \cap \cdots \cap A_{i_r}.$$

Then (11.2) can be written as

$$\Phi\left(\bigcup_{i=1}^m A_i\right) = \sum_{v \in S(m)} (-1)^{|v|-1} \Phi(A_v). \qquad (11.3)$$

11.2.3. LEMMA. *If* $\Phi : \mathcal{F} \to R$ *is fully additive on* $\mathcal{F}$, *then for any* $A_1, \ldots, A_m, B_1, \ldots, B_k \in \mathcal{F}$, *with* $\bigcup_{i=1}^m A_i = \bigcup_{j=1}^k B_j$,

$$\sum_{v \in S(m)} (-1)^{|v|-1} \Phi(A_v) = \sum_{\lambda \in S(k)} (-1)^{|\lambda|-1} \Phi(B_\lambda).$$

Proof. Let

$$\alpha := \sum_{v \in S(m)} (-1)^{|v|-1} \Phi(A_v), \quad \beta := \sum_{\lambda \in S(k)} (-1)^{|\lambda|-1} \Phi(B_\lambda),$$

and

$$\bigcup_{i=1}^m A_i = X = \bigcup_{j=1}^k B_j.$$

Then for every $v \in S(m)$ and $\lambda \in S(k)$,

$$A_v = X \cap A_v = \bigcup_{j=1}^k A_v \cap B_j \qquad (11.4)$$

and

$$B_\lambda = X \cap B_\lambda = \bigcup_{i=1}^{m} A_i \cap B_\lambda. \tag{11.5}$$

By (11.4) and (11.3) (full additivity of Φ),

$$\alpha = \sum_{\nu \in S(m)} (-1)^{|\nu|-1} \Phi\left(\bigcup_{j=1}^{k} (A_\nu \cap B_j)\right)$$

$$= \sum_{\nu \in S(m)} (-1)^{|\nu|-1} \sum_{\lambda \in S(k)} (-1)^{|\lambda|-1} \Phi(A_\nu \cap B_\lambda).$$

By (11.5) and (11.3),

$$\beta = \sum_{\lambda \in S(k)} (-1)^{|\lambda|-1} \Phi\left(\bigcup_{i=1}^{m} (A_i \cap B_\lambda)\right)$$

$$= \sum_{\lambda \in S(k)} (-1)^{|\lambda|-1} \sum_{\nu \in S(m)} (-1)^{|\nu|-1} \Phi(A_\nu \cap B_\lambda).$$

Hence $\alpha = \beta$. $\square$

In view of Lemma 11.2.3, every fully additive function $\Phi : \mathcal{F} \to \mathbb{R}$ can be extended over $\mathcal{U}(\mathcal{F})$ to the function Φ' defined by

$$\Phi'(X) := \sum_{\nu \in S(m)} (-1)^{|\nu|-1} \Phi(A_\nu) \tag{11.6}$$

for any $A_1, \ldots, A_m \in \mathcal{F}$ with $\bigcup_{i=1}^{m} A_i = X$.

Let us note that Lemma 11.2.3 is indispensable, because it guarantees that the value of Φ' for $X \in \mathcal{U}(\mathcal{F})$ is independent of the representation of X as a union of elements of $\mathcal{F}$.

The following is well known (compare with Exercise 11.3).

11.2.4. PROPOSITION. *The extension* Φ' *of a fully additive valuation* $\Phi : \mathcal{F} \to \mathbb{R}$, *defined by* (11.6), *is a valuation on* $\mathcal{U}(\mathcal{F})$.

(Proposition 11.2.4 is a part of Theorem 3.4.11 in [64]).

We shall prove that such an extension is fully additive:

11.2.5. PROPOSITION. *The function* Φ' *defined by* (11.6) *is fully additive on* $\mathcal{U}(\mathcal{F})$.

Proof. In view of 11.2.4, the assertion is true for $m = 2$. Let $m > 2$; we admit the inductive assumption for $m - 1$. Then

$$\Phi\left(\bigcup_{i=1}^{m} A_i\right) = \Phi\left(\bigcup_{i=1}^{m-1} A_i \cup A_m\right) = \Phi\left(\bigcup_{i=1}^{m-1} A_i\right) + \Phi(A_m) - \Phi\left(\bigcup_{i=1}^{m-1} (A_i \cap A_m)\right)$$

$$= \sum_{r=1}^{m-1} (-1)^{r-1} \sum_{i_1 < \cdots < i_r} \Phi(A_{i_1} \cap \cdots \cap A_{i_r}) + \Phi(A_m)$$

$$- \sum_{r=1}^{m-1} (-1)^{r-1} \sum_{i_1 < \cdots < i_r} \Phi(A_{i_1} \cap \cdots \cap A_{i_r} \cap A_m)$$

$$= \sum_{r=1}^{m} (-1)^{r-1} \sum_{i_1 < \cdots < i_r} \Phi(A_{i_1} \cap \cdots \cap A_{i_r}). \qquad \square$$

In Section 6.1, the Euler–Poincaré characteristic χ was introduced for geometric polyhedra. According to 6.1.5–6.1.7, since every compact convex subset of R^n is homeomorphic to a simplex, we define

$$\chi(A) := 1 \quad \text{for} \quad A \in \mathcal{K}^n \quad \text{and} \quad \chi(\emptyset) = 0. \tag{11.7}$$

Of course, every geometric polyhedron in R^n belongs to the convexity ring $\mathcal{U}^n$. Our next task is to extend the function χ defined on the family of geometric polyhedra and compact convex sets in R^n to a valuation $\chi' : \mathcal{U}^n \to R$. We shall need the following lemma. The proof is due to Groemer ([24]).

11.2.6. LEMMA. *The function $\chi | \mathcal{K}^n \cup \{\emptyset\}$ defined by (11.7) is fully additive on* $\mathcal{K}^n \cup \{\emptyset\}$.

Proof. Let $A_1, \ldots, A_m \in \mathcal{K}^n$ and $A = \bigcup_{i=1}^{m} A_i \in \mathcal{K}^n$ for some $m > 1$.
If $A_\nu \neq \emptyset$ for every $\nu \in S(m)$, i.e., all intersections of elements of the set $\{A_1, \ldots, A_m\}$ are nonempty, then the formula (11.2) can be written as

$$1 = \sum_{r=1}^{m} (-1)^{r-1} \binom{m}{r},$$

which is equivalent to

$$\sum_{r=0}^{m} (-1)^r \binom{m}{r} = 0;$$

but the last equality follows from the Newton formula.
Hence, for $m = 2$ the assertion is true, since $A_1 \cap A_2 \neq \emptyset$ as $A_1 \cup A_2$ is convex. Let $m > 2$; assume that the assertion is true for $m - 1$.

(a) If among $A_1, \ldots, A_m$ there is a pair of disjoint sets, then these sets can be separated by some closed half-spaces E', E'' (compare with 3.3.8). We may assume these two sets to be A_{m-1}, A_m; thus

$$A_{m-1} \subset E', \quad A_m \subset E'', \quad E' \cup E'' = R^n,$$

and $E' \cap E''$ is a hyperplane disjoint from $A_{m-1} \cup A_m$.

Let
$$A_i' = A_i \cap E', \quad A_i'' = A_i \cap E'',$$
and
$$A' := \bigcup A_i', \quad A'' := \bigcup A_i''.$$

By the inductive assumption,

$$\chi(A') = \sum_{r=1}^{m-1} (-1)^{r-1} \sum_{i_1 < \cdots < i_r} \chi(A_{i_1}' \cup \cdots \cup A_{i_r}')$$

and so for A'' and A_i''.

Let
$$E_0 := E' \cap E'', \quad A_0 := A' \cap A'' = A \cap E_0.$$

Since

$$A_0 = \left(\bigcup_{i=1}^{m} A_i \right) \cap E_0 = \bigcup_{i=1}^{m-2} A_i \cap E_0 \in \mathcal{K}^n,$$

from the inductive assumption it follows that

$$\chi(A_0) = \sum_{r=1}^{m-2} (-1)^{r-1} \sum_{i_1 < \cdots < i_r} \chi(A_{i_1} \cap \cdots \cap A_{i_r} \cap E_0),$$

whence

$$\chi(A) = \chi(A' \cup A'') = \chi(A') + \chi(A'') - \chi(A_0)$$
$$= \sum_{r=1}^{m} (-1)^{r-1} \sum_{i_1 < \cdots < i_r} \left(\chi(A_{i_1}' \cap \cdots \cap A_{i_r}') + \chi(A_{i_1}'' \cap \cdots \cap A_{i_r}'') \right)$$
$$= \sum_{r=1}^{m} (-1)^{r-1} \sum_{i_1 < \cdots < i_r} \chi(A_{i_1} \cap \cdots \cap A_{i_r}),$$

because

$$A_{i_1} \cap \cdots \cap A_{i_r} = (A_{i_1}' \cap \cdots \cap A_{i_r}') \cup (A_{i_1}'' \cap \cdots \cap A_{i_r}'').$$

(b) Assume that every two elements of the set $\{A_1, \ldots, A_m\}$ have a nonempty intersection:

$$A_i \cap A_j \neq \emptyset \quad \text{for} \quad i, j \in \{1, \ldots, m\}. \tag{11.8}$$

Let

$$k_0 := \max\{k \in \{1, \ldots, m\} \mid A_{i_1} \cap \cdots \cap A_{i_k} \neq \emptyset \quad \text{for every} \quad \{i_1, \ldots, i_k\}\}.$$

Of course, $k_0 \geq 2$. We have proved that the assertion is true for $k_0 = m$. Now let $k_0 < m$ and let, for example,

$$A_1 \cap \cdots \cap A_{k_0+1} = \emptyset. \tag{11.9}$$

We define $B_1, \ldots, B_m$:

$$B_i =: A_1 \cap \cdots \cap A_{k_0-1} \cap A_i.$$

Evidently, $B_i \in \mathcal{K}^n$ for $i = 1, \ldots, m$ and $B_{k_0} \cap B_{k_0+1} = \emptyset$ by (11.9). Thus we can use (a) for $B_1, \ldots, B_m$. But

$$\bigcup_{i=1}^{m} B_i = A \cap A_1 \cap \cdots \cap A_{k_0-1} = A_1 \cap \cdots \cap A_{k_0-1}$$

and

$$B_{i_1} \cap \cdots \cap B_{i_r} = (A_{i_1} \cap \cdots \cap A_{i_r}) \cap (A_1 \cap \cdots \cap A_{k_0-1}),$$

whence (by (a) for B_i) we obtain

$$\chi\left(\bigcup B_i\right) = \sum_{v}(-1)^{|v|-1}\chi((B_i)_v),$$

and this condition in terms of the sets $A_1, \ldots, A_m$ has the form

$$1 = \chi(A_1 \cap \cdots \cap A_{k_0-1})$$

$$= \sum_{r=1}^{m}(-1)^{r-1} \sum_{i_1 < \cdots < i_r} \chi((A_{i_1} \cap \cdots \cap A_{i_r}) \cap (A_1 \cap \cdots \cap A_{k_0-1}))$$

$$= \sum_{r=1}^{m}(-1)^{r-1} \sum (\chi(A_{i_1} \cap \cdots \cap A_{i_r}) + \chi(A_1 \cap \cdots \cap A_{k_0-1})$$

$$- \chi((A_{i_1} \cap \cdots \cap A_{i_r}) \cup (A_1 \cap \cdots \cap A_{k_0-1})).$$

Since the last two terms can be reduced (each of them equals 1), and $\chi(A) = 1$ (because $A \in \mathcal{K}^n$), it follows that

$$\chi(A) = \sum_{r=1}^{m}(-1)^{r-1} \sum_{i_1 < \cdots < i_r} \chi(A_{i_1} \cap \cdots \cap A_{i_r}). \qquad \square$$

In view of 11.2.4 and 11.2.6, the following formula defines the valuation χ' : $\mathcal{U}^n \to R$, which is an extension of the characteristic $\chi : \mathcal{K}^n \to R$: for any $A \in \mathcal{U}^n$, if $A = \bigcup_{i=1}^{m} A_i$, then

$$\chi'(A) := \sum_{v \in S(m)}(-1)^{|v|-1}\chi(A_v). \tag{11.10}$$

We shall now prove that χ' is also an extension of the Euler–Poincaré characteristic χ defined earlier for geometric polyhedra in R^n.

11.2.7. THEOREM. *For every geometric polyhedron P in $\mathbf{R}^n$,*

$$\chi'(P) = \chi(P).$$

Proof. Let $\mathcal{T}$ be a triangulation of the polyhedron P. We select the subset $\mathcal{A}$ of $\mathcal{T}$ consisting of the simplices that are not proper faces of another. Let $\mathcal{A} = \{A_1, \ldots, A_m\}$.

Since evidently $P = \bigcup \mathcal{A}$, and (by 11.2.3) the value of χ' is independent of the decomposition of a given set into a union of elements of $\mathcal{K}^n$, it suffices to prove that

$$\sum_{v \in S(m)} (-1)^{|v|-1} \chi(A_v) = \chi(P).$$

For $m = 1$ the set P is a simplex, whence this condition is satisfied. Let $m > 1$; assume that it is satisfied for $m - 1$.

Let $P' := \bigcup_{i=1}^{m-1} A_i$. Then $P = P' \cup A_m$ and

$$\chi(P) = \chi(P') + \chi(A_m) - \chi(P' \cap A_m).$$

An analogous formula holds for χ', because it is a valuation on $\mathcal{U}^n$:

$$\chi'(P) = \chi'(P') + \chi'(A_m) - \chi'(P' \cap A_m).$$

Since by the inductive assumption, the right-hand sides of these two formulas are equal, it follows that $\chi(P) = \chi'(P)$. $\qquad\square$

11.2.8. REMARK. To any finite set $\{A_1, \ldots, A_m\}$ in $\mathcal{K}^n$ a simplicial complex $\mathcal{N}(A_1, \ldots, A_m)$ can be assigned.

Let $S := \Delta(e_1, \ldots, e_m)$, where $e_i = (\delta_1^i, \ldots, \delta_m^i) \in \mathbf{R}^m$, and let $\mathcal{T}$ be the natural triangulation of the simplex S:

$$\mathcal{T} = \{\Delta(e_{i_1}, \ldots, e_{i_r}) \mid i_1, \ldots, i_r \in \{1, \ldots, m\}, r = 1, \ldots, m\}.$$

The complex $\mathcal{N}(A_1, \ldots, A_m)$ is a subcomplex of $\mathcal{T}$ defined as follows:

$$\mathcal{N}(A_1, \ldots, A_m)^{(0)} := S^{(0)} = \{e_1, \ldots, e_m\}, \tag{11.11}$$

and for every $r \in \{2, \ldots, m\}$,

$$\Delta(e_{i_1}, \ldots, e_{i_r}) \in \mathcal{N}(A_1, \ldots, A_m)^{(r)} :\Longleftrightarrow A_{i_1} \cap \cdots \cap A_{i_r} \neq \emptyset. \tag{11.12}$$

By analogy with the well-known topological notion of the nerve of a covering, the complex $\mathcal{N}(A_1, \ldots, A_m)$ is called the *nerve of* $\{A_1, \ldots, A_m\}$.

Directly from 6.1.3 combined with (11.11) and (11.12) it follows that

$$\chi'(A_1 \cup \cdots \cup A_m) = \chi(\mathcal{N}(A_1, \ldots, A_m)). \qquad\square$$

For the function χ' we use the same symbol χ and we call it the *Euler–Poincaré characteristic* .

Evidently, $\chi | \mathcal{K}^n = V_0$, whence χ is an extension on $\mathcal{U}^n$ of the basic functional V_0.

Let

$$V_0'(A) := \begin{cases} 0 & \text{if } A = \emptyset \\ \chi(A) & \text{if } \emptyset \neq A \in \mathcal{U}^n. \end{cases}$$

We shall now extend the remaining basic functionals, $V_1, \ldots, V_n$. We preserve the notation introduced in Chapter 9 (see 9.2.1).

11.2.9. DEFINITION. Let $n \geq 2$. For $i \in \{1, \ldots, n\}$, let

$$\beta_{n,i} := \frac{2}{i} \cdot \frac{\kappa_{i-1}}{\kappa_i \kappa_{n-1}},$$

and for $A \in \mathcal{U}^n$, let

$$V_i'(A) := \beta_{n,i} \int_{\mathcal{E}_A} V_{i-1}'(A \cap E) d\mu(E).$$

11.2.10. THEOREM. *The functionals* $V_0', \ldots, V_n'$ *are valuations on* $\mathcal{U}^n$ *and* $V_i'|\mathcal{K}^n = V_i$ *for* $i = 0, \ldots, n$.

Proof. We already know that theorem holds for $i = 0$. Let $i \geq 1$. By the Crofton Theorem 9.2.6,

$$V_i'(A) = V_i(A) \quad \text{for } A \in \mathcal{K}^n.$$

Evidently, for every i,

$$V_i'(\emptyset) = 0. \tag{11.13}$$

We prove by induction that V_i' is a valuation: for $i = 0$ it is a corollary of 11.2.4 combined with 11.2.6. Let $i \geq 1$; we admit the inductive assumption for $i - 1$. By Definition 11.2.9, condition (11.13), and the inductive assumption,

$$V_i'(A_1 \cup A_2) + V_i'(A_1 \cap A_2) = \beta_{n,i} \int_{\mathcal{E}^n} (V_{i-1}'(A_1 \cap E) + V_{i-1}'(A_2 \cap E)) \, d\mu(E)$$

$$= V_i'(A_1) + V_i'(A_2). \qquad \square$$

Directly from Definition 11.2.9 we infer that for the extended basic functionals the Crofton formulae hold.

12

Selectors for Convex Bodies

12.1 Symmetry centers

Centrally symmetric sets play a particular role in the geometry of R^n.

12.1.1. DEFINITION. Let A be a nonempty subset of R^n. A point p is a *symmetry center of* A if and only if $\sigma_p(A) = A$, i.e.,

$$x \in A \Longrightarrow \sigma_p(x) \in A.$$

Of course, some sets have many symmetry centers (compare Exercise 12.1). Let $C_0(A)$ be the set of all symmetry centers of A.

12.1.2. THEOREM. *For every nonempty convex subset A of R^n, the set $C_0(A)$ is a convex subset of A.*

Proof. We may assume that $C_0(A) \neq \emptyset$.

Let $p \in C_0(A)$. According to 12.1.1, if $x \in A$, then $\sigma_p(x) \in A$; thus $p = \frac{1}{2}(x + \sigma_p(x)) \in A$, because A is convex. Hence

$$C_0(A) \subset A.$$

Let $p_0, p_1 \in C_0(A)$ and $p_t := (1 - t)p_0 + tp_1$ for $t \in [0, 1]$.
For every $x \in A$,

$$\sigma_{p_t}(x) = 2p_t - x = (1 - t)(2p_0 - x) + t(2p_1 - x)$$

and $2p_i - x \in A$ for $i = 0, 1$; since A is convex, it follows that $\sigma_{p_t}(x) \in A$ and thus $p_t \in C_0(A)$. Hence $C_0(A)$ is convex. $\qquad \square$

12.1.3. THEOREM. *Every bounded subset of* R^n *has at most one symmetry center.*

Proof. We may assume $A \neq \emptyset$. Suppose that $p, q \in C_0(A)$ and $\alpha := \|p-q\| > 0$.

From 12.1.2 it follows that $p, q \in A$. Let $p_1 := p$ and

$$p_{2i} := \sigma_q(p_{2i-1}), \quad p_{2i+1} := \sigma_p(p_{2i}) \tag{12.1}$$

for $i \geq 1$.

Evidently, $p_j \in A$ for every $j \in N$. All the points of the sequence $(p_j)_{j \in N}$ belong to aff(p, q) in the following natural order (it means the order that corresponds to the order $\leq$ or $\geq$ in R by an isometry $R \to$ aff(p, q)):

$$\ldots, p_{2i+3}, p_{2i+1}, \ldots, p, q, p_2, \ldots, p_{2i+2}, p_{2i+4}, \ldots.$$

Hence by (12.1),

$$\|p_{2i} - p\| = 2i\alpha \tag{12.2}$$

(compare Exercise 12.2), and thus $\lim_{i \to \infty} \|p_{2i} - p\| = \infty$, contrary to the assumption that A is bounded. $\qquad \square$

Let $\mathcal{K}_1^n$ be the family of sets in $\mathcal{K}^n$ that have a symmetry center; for every $A \in \mathcal{K}_1^n$, the unique (by 12.1.3) symmetry center will be denoted by $c_0(A)$.

12.1.4. THEOREM. *For every $A \in \mathcal{K}_1^n$ and every affine automorphism f of R^n,*

$$c_0(f(A)) = f(c_0(A)).$$

Proof. For any $x, y \in R^n$,

$$f\left(\frac{x+y}{2}\right) = \frac{f(x) + f(y)}{2}. \tag{12.3}$$

Indeed, let $\bar{f}$ be the linear part of f:

$$\bar{f} := f(x) - f(0);$$

then

$$f\left(\frac{x+y}{2}\right) = f(0) + \bar{f}\left(\frac{x+y}{2}\right) = f(0) + \frac{\bar{f}(x) + \bar{f}(y)}{2} = \frac{f(x) + f(y)}{2}.$$

Since

$$p = c_0(A) \iff \forall x \in A \, \exists y \in A \quad \frac{x+y}{2} = p,$$

from (12.3) the assertion follows. $\qquad \square$

12.2 Selectors and multiselectors

12.2.1. DEFINITION. Let $\mathcal{F}$ be a nonempty family of convex subsets of R^n. A function $\phi : \mathcal{F} \to R^n$ is said to be a *selector* for $\mathcal{F}$ if

$$\forall A \in \mathcal{F} \quad \phi(A) \in A;$$

a function $\Phi : \mathcal{F} \to 2^{R^n}$ is called a *multiselector for* $\mathcal{F}$ if

$$\forall A \in \mathcal{F} \quad \emptyset \neq \Phi(A) \subset A.$$

Thus a selector chooses a point from each member of a given family, while a multiselector chooses a subset from each member.

12.2.2. EXAMPLE. Let $\mathcal{F}_1$ be the family of all convex subsets of R^n with at least one symmetry center, and let $\mathcal{F}_2$ be the subfamily of $\mathcal{F}_1$ consisting of bounded sets. In view of 12.1.2, the function $C_0 : \mathcal{F}_1 \to 2^{R^n}$ described in Section 12.1 is a multiselector for $\mathcal{F}_1$, while in view of 12.1.3, the function $c_0 : \mathcal{F}_2 \to R^n$ is a selector for $\mathcal{F}_2$.

12.2.3. DEFINITION. A selector $\phi : \mathcal{F} \to R^n$ is *equivariant under a map* $f : R^n \to R^n$ if
$$\forall A \in \mathcal{F} \quad \phi(f(A)) = f(\phi(A)).$$

12.2.4. EXAMPLE. The selector $c_0 : \mathcal{K}_1^n \to R^n$ is equivariant under affine automorphisms (compare 12.1.4).

As we shall see, for compact, convex, centrally symmetric sets, c_0 is the only selector equivariant under all the isometries:

12.2.5. THEOREM. *If $\phi : \mathcal{K}^n \to R^n$ is equivariant under the isometries, then*

$$\phi|\mathcal{K}_1^n = c_0.$$

Proof. Let $A \in \mathcal{K}_1^n$ and $p = c_0(A)$. Then by assumption,

$$\sigma_p(\phi(A)) = \phi(\sigma_p(A)) = \phi(A),$$

whence $\phi(A)$ is the fixed point of the central symmetry σ_p. Thus

$$\phi(A) = p. \qquad \square$$

Theorem 12.2.5 is a particular case of the following one (compare Exercise 12.4):

12.2.6. THEOREM. *If a selector $\phi : \mathcal{K}^n \to R^n$ is equivariant under the isometries, then for every $A \in \mathcal{K}^n$ and every affine subspace E*

$$\sigma_E(A) = A \implies \phi(A) \in E.$$

12.3 Centers of gravity

The notion of a center of gravity has its roots in physics. For a system of "material points," i.e., points with weights, one looks for the point for which the system is in the equilibrium state.

In terms of the geometry of R^n this can be described as follows.

12.3.1. DEFINITION. Let $k \in N$. Let $p_1, \ldots, p_k \in R^n$, $\alpha_1, \ldots, \alpha_k \geq 0$, and $\sum_{i=1}^{k} \alpha_i \neq 0$. A point x_0 is the *center of gravity of* $(p_1, \ldots, p_k)$ *with the system of weights* $(\alpha_1, \ldots, \alpha_k)$ if

$$\sum_{i=1}^{k} \alpha_i (p_i - x_0) = 0. \tag{12.4}$$

12.3.2. THEOREM. *Let* $p_1, \ldots, p_k \in R^n$. *For every sequence* $(\alpha_1, \ldots, \alpha_k)$ *of nonnegative numbers, at least one different from 0, there exists a unique center of gravity* x_0 *of* $(p_1, \ldots, p_k)$ *with the system of weights* $(\alpha_1, \ldots, \alpha_k)$:

$$x_0 := \frac{\sum_{i=1}^{k} \alpha_i p_i}{\sum_{i=1}^{k} \alpha_i}. \tag{12.5}$$

Proof. Let us first show that the point x_0 defined by (12.5) is a center of gravity of $(p_1, \ldots, p_k)$ with the system of weights $(\alpha_1, \ldots, \alpha_k)$. Indeed,

$$\sum_{j=1}^{k} \alpha_j (p_j - x_0) = \sum_{j=1}^{k} \alpha_j p_j - \frac{1}{\sum_{i=1}^{k} \alpha_i} \cdot \sum_{j=1}^{k} \alpha_j \sum_{i=1}^{k} \alpha_i p_i = 0,$$

whence condition (12.4) is satisfied.

Suppose now that there is another center of gravity y_0 of $(p_1, \ldots, p_k)$ with the same system of weights. Then

$$\sum_{j=1}^{k} \alpha_j (p_j - x_0) = 0 = \sum_{j=1}^{k} \alpha_j (p_j - y_0);$$

thus $\sum_{j=1}^{k} \alpha_j (x_0 - y_0) = 0$. Hence $x_0 = y_0$, because $\sum_{j=1}^{k} \alpha_j \neq 0$. □

This elementary notion of a center of gravity has a counterpart for an arbitrary measure space $(X, \mathcal{A}, \mu)$, in particular for $(R^n, \mathcal{B}(R^n), \mu)$ with μ being an arbitrary Borel measure:

12.3.3. DEFINITION. A point x_0 is the *center of gravity of a set* $A \in \mathcal{B}(R^n)$ *with respect to* μ *if*

$$\int_A (x - x_0) \, d\mu(x) = 0.$$

12.3.4. EXAMPLE. If A is finite, $A = \{p_1, \ldots, p_k\}$, μ is a nonzero measure, and $\mu(\{p_i\}) = \alpha_i$ for $i = 1, \ldots, k$, then the center of gravity of the set A with respect to μ is the center of gravity of the sequence $(p_1, \ldots, p_k)$ with the system of weights $(\alpha_1, \ldots, \alpha_k)$.

The following theorem is a generalization of 12.3.2.

12.3.5. THEOREM. *Let μ be a Borel measure. For every $A \in \mathcal{B}(\mathbb{R}^n)$ with $0 < \mu(A) < \infty$, there is a unique center of gravity x_0 of A with respect to μ:*

$$x_0 := \frac{\int_A x \, d\mu(x)}{\mu(A)}. \tag{12.6}$$

Proof. The point x_0 defined by (12.6) is a center of gravity of A with respect to μ. Indeed,

$$\int_A (x - x_0) \, d\mu(x) = \frac{1}{\mu(A)} \cdot \left(\mu(A) \int_A x \, d\mu(x) - \int_A \int_A x \, d\mu(x) \, d\mu(y) \right)$$

$$= \frac{1}{\mu(A)} \left(\int_A x \, d\mu(x) \right) \left(\mu(A) - \int_A d\mu(y) \right) = 0.$$

Suppose there is another center of gravity y_0. Then

$$\int_A (x - x_0) \, d\mu(x) = 0 = \int_A (x - y_0) \, d\mu(x),$$

whence $\int_A (x_0 - y_0) \, d\mu(x) = 0$. Thus $x_0 = y_0$, because $0 < \mu(A) < \infty$.   □

12.3.6. DEFINITION. For any set $A \in \mathcal{B}(\mathbb{R}^n)$ and any Borel measure μ with $0 < \mu(A) < \infty$, let $c_\mu(A)$ be the center of gravity of A with respect to μ.

12.3.7. LEMMA. *If a Borel measure μ on $\mathbb{R}^n$ is invariant under an isometry $f : \mathbb{R}^n \to \mathbb{R}^n$, then for every $A \in \mathcal{B}(\mathbb{R}^n)$ with $0 < \mu(A) < \infty$,*

$$c_\mu(f(A)) = f(c_\mu(A)).$$

The proof is left to the reader (Exercise 12.6).

12.3.8. THEOREM. *If a Borel measure μ is invariant under isometries of $\mathbb{R}^n$, then $c_\mu | \mathcal{K}_0^n$ is a selector for $\mathcal{K}_0^n$. If $\mu = \lambda_n$, this selector is equivariant under the affine automorphisms.*

Proof. We are going to show that $c_\mu(A) \in A$ for every $A \in \mathcal{K}_0^n$.

Let $a = c_\mu(A)$. Suppose that $a \notin A$. Then there exists a hyperplane H that separates a from A.

Let $\alpha := \frac{1}{2}\varrho(a, A)$. In view of 12.3.7, we may assume that

$$H = \{(x_1, \ldots, x_n) \in \mathbb{R}^n \mid x_n = \alpha\}, \quad a = 0,$$

and
$$A \subset \{(x_1, \ldots, x_n) \in \mathbf{R}^n \mid x_n > \alpha\}.$$

Then
$$\int_A x_n d\mu(x) > \int_A \alpha d\mu(x) = \alpha\mu(A) > 0.$$

Since
$$\int_A x\, d\mu(x) = (y_1, \ldots, y_n),$$

where
$$y_i = \int_A x_i d\mu(x),$$

by (12.6) it follows that
$$a \cdot \mu(A) \neq 0,$$

contrary to the assumption $a = 0$.

Hence $c_\mu | \mathcal{K}_0^n$ is a selector. It remains to prove that for $\mu = \lambda_n$, the Lebesgue measure, this selector is equivariant under any affine automorphism f of $\mathbf{R}^n$. For translations it is obvious; thus we may assume f to be a linear automorphism. Then
$$V_n(f(A)) = |\det f| V_n(A),$$

and
$$\int_{f(A)} y\, d\lambda_n(y) = |\det f| f\left(\int_A x\, d\lambda_n(x)\right).$$

Thus
$$c_{\lambda_n}(f(A)) = \frac{1}{|\det f| V_n(A)} \int_{f(A)} y\, d\lambda_n(y)$$
$$= \frac{1}{V_n(A)} f\left(\int_A x\, d\lambda_n(x)\right) = f c_{\lambda_n}(A). \qquad \square$$

12.3.9. EXAMPLE. Let $A = \Delta(a_0, \ldots, a_n)$. Then
$$c_{\lambda_n}(A) = c\left(a_0, \ldots, a_n; \frac{1}{n}, \ldots, \frac{1}{n}\right);$$

thus $c_{\lambda_n}(A)$ is the center of gravity of the sequence of vertices of the simplex A, with equal weights (compare Theorem 12.3.2 and Exercise 12.5). For a regular simplex this follows from 12.2.6, because this point is the point of intersection of all the hyperplanes of symmetry of such a simplex. Since every n-dimensional simplex is the image of a regular one under an affine automorphism, for an arbitrary simplex it suffices to apply 12.3.8.

The center of gravity c_{λ_n} is also called the *centroid*.

The next example concerns another center of gravity.

12.3.10. EXAMPLE. (See [64], p. 305.) For every $A \in \mathcal{K}_0^n$ and $i \in \{0, \ldots, n\}$, let $p_i(A)$ be the center of gravity of A with respect to the $(n - i)$-dimensional curvature measure of A:

$$p_i(A) := c_{\Phi_{n-i}(A,\cdot)}(A).$$

In particular, $p_0(A)$ is the centroid of A, i.e.,

$$p_0(A) = c_{\lambda_n}(A)$$

(compare (10.6)).

Since by 10.1.12, the curvature measures are invariant under isometries, from 12.3.8 it follows that for each i the map p_i is a selector.

The following theorem, which describes the position of the centroid of A, has been well known for many years; an idea of a proof was suggested in [10]. The proof given below can be found in [53].

12.3.11. THEOREM. *Let* $u \in S^{n-1}$.
(i) *For every* $A \in \mathcal{K}_0^n$,

$$b(A, u) \le (n + 1)\varrho(c_{\lambda_n}(A), H(A, u)).$$

(ii) *If in particular,* A *is a cone with base contained in* $H(A, u)$, *then the inequality is an equality.*

Proof. Let $A \in \mathcal{K}_0^n$ and $x_0 := c_{\lambda_n}(A)$. For every $t \in \mathbb{R}$, let

$$H_t = \{x \in \mathbb{R}^n \mid x \circ u = t\}.$$

If $u = e_n$ and $0 \in A$, then

$$u \circ x_0 = \frac{1}{V_n(A)} \int_{-h(A,-u)}^{h(A,u)} \int_{A \cap H_t} t \, d\lambda_{n-1}(x_1, \ldots, x_{n-1}) \, dt$$

$$= \frac{1}{V_n(A)} \int_{-h(A,-u)}^{h(A,u)} t V_{n-1}(A \cap H_t) \, dt.$$

(ii): Let A be a cone with base B and vertex 0. Then

$$H_0 = H(A, -u) \quad \text{and} \quad H_{b(A,u)} = H(A, u).$$

Using two elementary properties of cones,

$$\frac{V_{n-1}(A \cap H_t)}{V_{n-1}(B)} = \left(\frac{t}{b(A, u)}\right)^{n-1}$$

and

$$V_n(A) = \frac{1}{n} V_{n-1}(B) b(A, u),$$

we obtain

$$\varrho(x_0, H(A, -u)) = u \circ x_0 = \frac{n}{b(A, u)^n} \int_0^{b(A,u)} t^n dt = \frac{n}{n+1} b(A, u).$$

Thus

$$\varrho(x_0, H(A, u)) = \frac{1}{n+1} b(A, u),$$

which proves (ii).

(i): Now let A be an arbitrary convex body. As above, we may assume that $u = e_n$ and $x_0 = 0$.

Let D be a cone with base $B \subset H(A, u)$, with vertex in $A \cap H(A, -u)$, and with $D \cap H_0 = A \cap H_0$. Let $y_0 := c_{\lambda_n}(D)$.

In view of (ii), it suffices to prove that

$$\varrho(x_0, H(A, u)) \geq \varrho(y_0, H(A, u)),$$

or equivalently,

$$u \circ y_0 \geq 0.$$

Indeed, it is easy to see that

$$t V_{n-1}(D \cap H_t) \geq t V_{n-1}(A \cap H_t)$$

for every $t \in [-h(A, -u), h(A, u)]$; thus

$$u \circ y_0 \geq \frac{V_n(A)}{V_n(D)} u \circ x_0 = 0. \qquad \square$$

We complete this section with a useful formula for the centroid of a convex body.

12.3.12. THEOREM. *For every* $A \in \mathcal{K}_0^n$,

$$c_{\lambda_n}(A) = \frac{1}{n+1} \int_{S^{n-1}} u \rho_A(u)^{n+1} d\sigma(u).$$

Proof. Since the centroid map is equivariant under translations, we may assume that $A \subset \{(x_1, \ldots, x_n) \in \mathbb{R}^n \mid x_i \geq 0 \text{ for } i = 1, \ldots, n\}$.

Thus the desired formula follows from Theorem 7.3.5 by simple calculation.$\square$

12.4 The Steiner point

An interesting example of a selector for $\mathcal{K}^n$ is the so-called Steiner point. The history of this notion can be found in [60].

12.4.1. DEFINITION. For every $A \in \mathcal{K}_0^n$, let

$$s(A) := \frac{1}{\kappa_n} \int_{S^{n-1}} u h_A(u) d\sigma(u).$$

The point $s(A)$ is the *Steiner point of A*.

We mention without proof the following result (compare [64] (5.4.12) and [60]).

12.4.2. THEOREM. *Let $A \in \mathcal{K}_0^n$. The Steiner point of A is the center of gravity of A with respect to the curvature measure $\Phi_0(A, \cdot)$:*

$$s(A) = c_{\Phi_0(A, \cdot)}(A).$$

From 12.2.4 combined with definition of the function p_i (see 12.3.10), it follows that

$$s = p_n. \tag{12.7}$$

Thus by 12.3.8, the map s is a selector for $\mathcal{K}_0^n$.

The notion of the Steiner point has a simple geometric interpretation for polytopes:

12.4.3. THEOREM. *Let $P \in \mathcal{P}_0^n$ and let $\{a_1, \ldots, a_r\}$ be the set of vertices of P. Then*

$$s(P) = \sum_{i=1}^{r} \gamma_0(P, \{a_i\}) \cdot a_i$$

(compare Definition 10.1.10).

Proof. Since the measure $\Phi_0(P, \cdot)$ is concentrated on the set of vertices of P, it suffices to apply 10.1.11 and 12.4.2.   □

The following theorem describes important properties of $s : \mathcal{K}^n \to \mathbb{R}^n$ ([60]).

12.4.4. THEOREM. *The selector s is Minkowski additive, equivariant under similarities, and continuous.*

Proof. Additivity follows from Theorem 3.4.4 concerning support functions; the proof of continuity and equivariance under similarities is an easy exercise (Exercise 12.8).   □

The following Schneider theorem ([60] Theorem 1) gives a solution of a modification of a problem posed by Grünbaum in [27]. We mention it without proof.

12.4.5. THEOREM. *If a function $\phi : \mathcal{K}^n \to \mathbb{R}^n$ is Minkowski additive, equivariant under the proper isometries, and continuous, then $\phi = s$.*

Let us note that in view of 12.4.4, the conditions in Theorem 12.4.5 sufficient for a function ϕ to coincide with s are also necessary. Thus they characterize the selector s.

Moreover, as a consequence of these two theorems we obtain the following corollary.

12.4.6. COROLLARY. *For every $\phi : \mathcal{K}^n \rightarrow \mathbf{R}^n$, the following conditions are equivalent:*

(i) *ϕ is additive, equivariant under similarities, and continuous,*

(ii) *ϕ is additive, equivariant under proper isometries, and continuous,*

(iii) *$\phi = s$.*

The selector s is not only continuous but is Lipschitz continuous (compare [64] (1.8.3) and Note 15 p. 61):

12.4.7. THEOREM. (see [56]) *For every A, $B \in \mathcal{K}^n$,*

$$\|s(A) - s(B)\| \leq \frac{2\kappa_{n-1}}{\kappa_n} \varrho_H(A, B).$$

This estimate is optimal.

12.5 Center of the minimal ring

The notion of the minimal ring containing the boundary of a convex body in $\mathbf{R}^n$ appeared in the literature in 1924 for $n = 2$ ([8]). Its short history can be found in the paper by Imre Bárány (see [2]).

For every $A \in \mathcal{K}_0^n$ and $x \in A$, let

$$R_A(x) := \inf\{\alpha > 0 \mid B(x, \alpha) \supset A\} \tag{12.8}$$

and

$$r_A(x) := \sup\{\alpha > 0 \mid B(x, \alpha) \subset A\}. \tag{12.9}$$

12.5.1. LEMMA. (i) *The function $R_A : A \rightarrow \mathbf{R}$ is convex.*

(ii) *If for distinct $x_0, x_1 \in A$,*

$$R_A\left(\frac{x_0 + x_1}{2}\right) = \frac{1}{2}(R_A(x_0) + R_A(x_1)),$$

then there is a unique point $p \in \mathrm{bd}\,A \cap \mathrm{aff}(x_0, x_1)$ such that

$$\left\|p - \frac{x_0 + x_1}{2}\right\| = R_A\left(\frac{x_0 + x_1}{2}\right) \quad \text{and} \quad \|p - x_i\| = R_A(x_i) \text{ for } i = 0, 1.$$

(iii) *The function $r_A : A \rightarrow \mathbf{R}$ is concave.*

Proof. (i): Let $x_0, x_1 \in A$ and $x_t = (1 - t)x_0 + tx_1$ for $t \in [0; 1]$. We have to prove that

$$R_A(x_t) \leq (1 - t)R_A(x_0) + tR_A(x_1). \tag{12.10}$$

By definition of R_A (see (12.8)), there exists $p \in \mathrm{bd}\,A$ such that

$$R_A(x_t) = \|x_t - p\|. \tag{12.11}$$

Since $A \subset B(x_i, R_A(x_i))$ for $i = 0, 1$, it follows that $p \in B(x_i, R_A(x_i))$ and thus $\|p - x_i\| \le R_A(x_i)$ for $i = 0, 1$. Hence by (12.11),

$$R_A(x_t) = \|(1 - t)(x_0 - p) + t(x_1 - p)\| \le (1 - t)\|x_0 - p\| + t\|x_1 - p\|$$
$$\le (1 - t)R_A(x_0) + tR_A(x_1),$$

i.e., (12.10) holds.

(ii): Evidently, the equality in (12.10) occurs if and only if there are only equalities in the last formula, i.e.,

$$\|(1 - t)(x_0 - p) + t(x_1 - p)\| = (1 - t)\|x_0 - p\| + t\|x_1 - p\|$$
$$= (1 - t)R_A(x_0) + tR_A(x_1).$$

Thus $p \in \mathrm{aff}(x_0, x_1)$ and $\|p - x_i\| = R_A(x_i)$ for $i = 0, 1$. This implies the uniqueness of p.

(iii): We have to prove that

$$r_A(x_t) \ge (1 - t)r_A(x_0) + tr_A(x_1) \tag{12.12}$$

for every $x_0, x_1 \in A$.

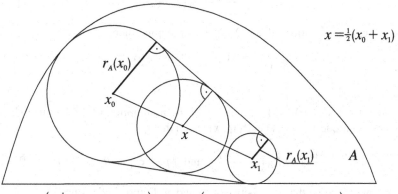

$$B\left(x, \tfrac{1}{2}(r_A(x_0) + r_A(x_1))\right) \subset \mathrm{conv}\left(B(x_0, r_A(x_0)) \cup B(x_1, r_A(x_1))\right)$$

Figure 12.1.

Obviously,
$$B(x_0, r_A(x_0)) \cup B(x_1, r_A(x_1)) \subset A.$$

Let us observe that (Figure 12.1)

$$B(x_t, (1 - t)r_A(x_0) + tr_A(x_1)) \subset \mathrm{conv}(B(x_0, r_A(x_0)) \cup B(x_1, r_A(x_1))),$$

because for any convex X, Y,

$$\text{conv}(X \cup Y) = \bigcup_{t \in [0; 1]} (1 - t)X + tY$$

(compare Exercise 3.9). Thus

$$B(x_t, (1 - t)r_A(x_0) + t r_A(x_1)) \subset A,$$

which yields (12.12).   □

12.5.2. THEOREM (I. Bárány [2]). *For every $A \in \mathcal{K}_0^n$ there is a unique mini-mizer x_0 of the function $R_A - r_A : A \to$ R.*

Proof. By Lemma 12.5.1, the function R_A is convex and r_A is concave; hence $R_A - r_A : A \to$ R is convex. The functions R_A and r_A can be extended over R^n to a convex function $\bar{R}_A$ and concave $\bar{r}_A$ (see [2] and [46]). Since $\bar{R}_A - \bar{r}_A : \mathrm{R}^n \to$ R is convex, it follows that it is continuous (compare 1.5.1 in [64]). Thus $R_A - r_A$, as a continuous function on a compact set A, attains its lower bound at some point x_0 of A. It remains to prove the uniqueness. Suppose that there exist two minimizers, x_0 and x_1, and let

$$\alpha := R_A(x_i) - r_A(x_i) \quad \text{for} \quad i = 0, 1.$$

We shall show that

$$R_A\left(\frac{x_0 + x_1}{2}\right) - r_A\left(\frac{x_0 + x_1}{2}\right) < \alpha. \tag{12.13}$$

Suppose, to the contrary, that (12.13) does not hold; then

$$R_A\left(\frac{x_0 + x_1}{2}\right) - r_A\left(\frac{x_0 + x_1}{2}\right) = \alpha,$$

because $R_A - r_A$ is convex by 12.5.1 (i).

Then by 12.5.1 (i), (iii), x_0, x_1 satisfy the assumption of 12.5.1 (ii), whence there is a unique $p \in \mathrm{bd} A \cap \mathrm{aff}(x_0, x_1)$ such that

$$\left\| p - \frac{(x_0 + x_1)}{2} \right\| = R_A\left(\frac{x_0 + x_1}{2}\right) \quad \text{and} \quad \| p - x_i \| = R_A(x_i) \text{ for } i = 0, 1.$$

We may assume that x_1 is between x_0 and p (Figure 12.2); since

$$r_A(x_0) - r_A(x_1) = R_A(x_0) - R_A(x_1),$$

it follows that $B(x_1, r_A(x_1)) \subset B(x_0, r_A(x_0))$, and thus $\|x_0 - x_1\| = r_A(x_0) - r_A(x_1)$. Hence there is exactly one point $q \in \mathrm{bd} A \cap \Delta(x_1, p)$ with $\|q - x_1\| = r_A(x_1)$. But $\text{conv}(B(x_0, r_A(x_0)) \cup \{p\})$ is a subset of A and q belongs to the interior of this convex hull; thus $q \in \mathrm{int} A$, a contradiction.

So, we have proved the inequality (12.13). From this inequality we infer that

$$\inf_{x \in A} (R_A(x) - r_A(x)) < \alpha,$$

contrary to the assumption.   □

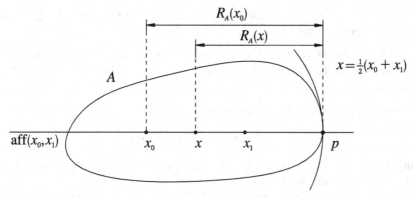

Figure 12.2. The figure illustrates an impossible situation; evidently $\|p - x_1\| \neq R_A(x_1)$.

Let $c(A)$ be the unique point x_0 whose existence is ensured by Theorem 12.5.2, and let

$$R(A) := R_A(c(A)) \quad \text{and} \quad r(A) = r_A(c(A)). \tag{12.14}$$

Let us consider the set $\operatorname{ring} A$ obtained from the ball $B(c(A), R(A))$ circumscribed about A by deleting the interior of the ball $B(c(A), r(A))$ inscribed in A:

$$\operatorname{ring} A := B(c(A), R(A)) \setminus \operatorname{int} B(c(A), r(A)). \tag{12.15}$$

The set $\operatorname{ring} A$ is called the *minimal ring of A*, and the point $c(A)$ is called the *center of the minimal ring*.

Of course, the function $c : \mathcal{K}_0^n \to \mathbf{R}^n$ is a selector for $\mathcal{K}_0^n$.

Directly from the definition of center of the minimal ring we obtain (compare Exercise 2.4)

12.5.3. PROPOSITION. *The selector c is equivariant under similarities.*

The following theorem is a consequence of 12.5.3 combined with 12.2.6.

12.5.4. THEOREM. *If a convex body A is symmetric with respect to an affine subspace E, then $c(A) \in E$.*

In particular, we obtain the following.

12.5.5. COROLLARY. *If a convex body A is centrally symmetric, then $c(A)$ is the center of symmetry.*

I. Bárány in [2] gave a criterion that allows one to check whether a point of A is the center of the minimal ring. We mention it without proof.

12.5.6. THEOREM ([2] Theorem 2). *Let $A \in \mathcal{K}_0^n$ and $x_0 \in \mathbf{R}^n$. A point x_0 is the center of the minimal ring of A if and only if there exist $p_1, \ldots, p_k \in \operatorname{bd} A \cap \operatorname{bd} B(x_0, R_A(x_0))$ and $q_1, \ldots, q_l \in \operatorname{bd} A \cap \operatorname{bd} B(x_0, r_A(x_0))$ such that*

$$\operatorname{conv}\left\{ \frac{p_i - x_0}{R_A(x_0)} \mid i = 1, \ldots, k \right\} \cap \operatorname{conv}\left\{ \frac{q_j - x_0}{r_A(x_0)} \mid j = 1, \ldots, l \right\} \neq \emptyset. \tag{12.16}$$

Applying 12.5.6, B. Zdrodowski found the center of the minimal ring for an arbitrary triangle in R^2:

12.5.7. THEOREM ([68]). *For any triangle T in R^2, the point $c(T)$ is the intersection point of the bisectrix of the longest side and the bisectrix of the smallest angle of T.*

Proof. Let $T = \Delta(a, b, c)$. Assume that

$$\|a - b\| \geq \max\{\|a - c\|, \|b - c\|\}$$

and that the measure of the angle at the vertex a is not greater than the measures of the two remaining angles.

Let x_0 be the point of intersection of the bisectrix of $\Delta(a, b)$ and the bisectrix of the angle at a.

We define p_1, p_2, q_1, q_2:

$$p_1 := a, \quad p_2 := b, \quad q_1 := \xi_{\mathrm{aff}(a,b)}(x_0), \quad q_2 := \xi_{\mathrm{aff}(a,c)}(x_0).$$

It suffices to show that these points satisfy condition (12.16). To this end let us observe that

$$R_T(x_0) = \|p_i - x_0\| \quad \text{and} \quad r_T(x_0) = \|q_j - x_0\|.$$

Without any loss of generality we may assume that $R_T(x_0) = 1$ and $x_0 = 0$ (we use Exercise 12.7). Then condition (12.16) reads as

$$\Delta(a, b) \cap \Delta\left(\frac{q_1}{r_A(0)}, \frac{q_2}{r_A(0)}\right) \neq \emptyset,$$

so evidently it is satisfied (Figure 12.3). □

The selector c is continuous but is not additive ([46]):

12.5.8. THEOREM. *Let $A, A_k \in \mathcal{K}_0^n$ for $k \in N$. Then*

$$\lim_H A_k = A \implies \lim c(A_k) = c(A).$$

12.5.9. THEOREM. *The function $c : \mathcal{K}_0^n \to R^n$ is not Minkowski additive.*

Proof. Suppose the selector c is additive. Since by 12.5.3 it is equivariant under isometries, and by 12.5.8 is continuous, in view of Theorem 12.4.5 it coincides with the Steiner point map: for every $A \in \mathcal{K}_0^n$,

$$c(A) = s(A).$$

We shall prove the opposite. Let $n = 2$ and let $A := \Delta(a, b, c)$, where $\angle(acb) = \frac{\pi}{2}$, $\|a - c\| = \|b - c\|$, and $c(A) = 0$. By 12.5.7 the point $c(A)$ is the center of the circle inscribed in A; it is easy to check that

$$\|c\| = \frac{\sqrt{2}}{2}\|a + b\|. \tag{12.17}$$

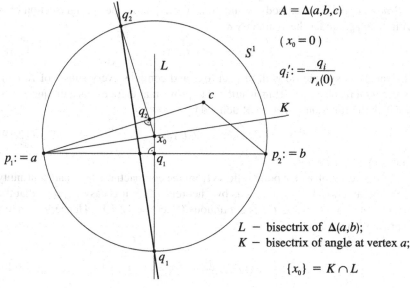

$$A = \Delta(a,b,c)$$

$$(x_0 = 0)$$

$$q_i' := \frac{q_i}{r_A(0)}$$

L — bisectrix of $\Delta(a,b)$;
K — bisectrix of angle at vertex a;

$$\{x_0\} = K \cap L$$

Figure 12.3.

On the other hand, by Theorem 12.4.3, $\frac{3\pi}{4}(a+b) + \frac{\pi}{2}c = 0$, whence

$$\|c\| = \frac{3}{2}\|a+b\|,$$

contrary to (12.17).   □

It was recently proved by I. Herburt that the center c of the minimal ring is not Lipschitz continuous (see [32]).

Finally, let us mention the following application of minimal rings (see [6]).

12.5.10. THEOREM. *For every* $A \in \mathcal{K}_0^n$, *there is a unique ball nearest to* A *with respect to* ϱ_H. *It has center* $c(A)$ *and radius* $\frac{1}{2}(R(A) + r(A))$.

The particular case of this result for $n = 2$ was already proved by H. Lebesgue in 1921 (see [9], p.52).

12.6 Pseudocenter. G-pseudocenters

In 1950 I. Fáry and L. Rédei proved (see [17]) that for every convex body A in R^n there is a unique centrally symmetric convex body contained in A with maximal volume:

12.6.1. THEOREM. *For every* $A \in \mathcal{K}_0^n$ *there is a unique* $C \in \mathcal{K}_1^n$ *such that*

$$C \subset A \quad \text{and} \quad V_n(C) \geq V_n(X) \quad \text{for every } X \in \mathcal{K}_1^n \text{ with } X \subset A.$$

Proof. For any convex body A and point $x \in A$, let A_x be the intersection of A with its image under the symmetry σ_x:

$$A_x := A \cap \sigma_x(A). \tag{12.18}$$

The set A_x is symmetric with respect to x and contains every subset of A symmetric with respect to x. Thus it suffices to prove that there exists a unique $x \in A$ at which the function $\psi : A \to \mathbb{R}$ defined by

$$\psi(x) := V_n(A_x) \tag{12.19}$$

attains its lower bound.

The existence of such a point follows from the compactness of A and continuity of ψ (the function V_n is continuous by Theorem 7.3.3; it is easy to prove that the function $A \ni x \mapsto A_x \in \mathcal{K}^n$ is continuous (Exercise 12.10)). Hence, it remains to prove the uniqueness.

Suppose $x_1 \neq x_2$ and

$$\psi(x_1) = \psi(x_2) \geq \psi(x) \quad \text{for every } x \in A. \tag{12.20}$$

Let $x_0 = \frac{1}{2}(x_1 + x_2)$. Then

$$A_{x_0} \supset \frac{1}{2}(A_{x_1} + A_{x_2}) \tag{12.21}$$

(Exercise 12.11). Thus by 5.2.2 (the Brunn–Minkowski inequality) combined with (12.19) and (12.20),

$$\psi(x_0) \geq \frac{1}{2^n}(\psi(x_1)^{\frac{1}{n}} + \psi(x_2)^{\frac{1}{n}})^n \geq \psi(x_0).$$

Hence in the last formula, both inequalities are equalities, which corresponds to the equality in the Brunn–Minkowski inequality for A_{x_1} and A_{x_2}. By Theorem 6.1.1 in [64], the set A_{x_2} is the image of A_{x_1} under a homothety:

$$A_{x_2} = \lambda A_{x_1} + v.$$

But $\lambda = 1$, because the volumes of these sets are equal, and $v \neq 0$, because $x_1 \neq x_2$ (Exercise 12.12).

Let $B := \mathrm{conv}(A_{x_1} \cup A_{x_2})$. Obviously, $B \in \mathcal{K}_1^n$, $B \subset A$, and $V_n(A_{c_0(B)}) \geq V_n(B) > V_n(A_{\hat{x}_i})$ for $i = 1, 2$, contrary to (12.20). $\qquad\square$

The unique convex body $C \in \mathcal{K}_1^n$ whose existence is ensured by Theorem 12.6.1 will be called the *symmetric kernel of A*; the symmetry center of C will be called the *pseudocenter of A*.[1]

In view of 12.6.1 and 12.1.3, every convex body A in $\mathbb{R}^n$ has exactly one pseudocenter; we denote it by $p(A)$. By 12.1.2, $p(A) \in A$. Thus we have the following.

---

[1] Our terminology differs from that in [17].

12.6.2. PROPOSITION. *The function* $p : \mathcal{K}_0^n \to \mathbb{R}^n$ *is a selector for* $\mathcal{K}_0^n$.

The following proposition is easy to prove (see Exercise 12.14):

12.6.3. PROPOSITION. *The selector* p *is equivariant under affine automorphisms of* $\mathbb{R}^n$.

12.6.4. EXAMPLE. The pseudocenter of any simplex A coincides with its centroid; thus according to 12.3.10,

$$p(A) = p_0(A).$$

Indeed, by the affine equivariance of the centroid and of the pseudocenter, we may assume A to be regular. Then by 12.2.6, the assertion follows from the equivariance of both selectors under isometries.

Let us observe that the problem of existence of a centrally symmetric convex body with the maximal volume contained in $A \in \mathcal{K}_0^n$ is a particular case of the following one, which was solved in [52], Theorem 3.8 (see Theorem 12.6.7 below).

12.6.5. PROBLEM. Let G be a subgroup of $O(n)$. Is it true that for every $A \in \mathcal{K}_0^n$ there is a unique convex body C contained in A with maximal volume, invariant under $\tau G \tau^{-1}$ for some $\tau \in \mathrm{Tr}$?

Indeed, Theorem 12.6.2 gives a positive answer to this question for $G = \{\mathrm{id}_{\mathbb{R}^n}, \sigma_0\}$, because C is symmetric with respect to x if and only if $\sigma_x(C) = C$ (recall that $\sigma_x = \tau_x \sigma_0 \tau_x^{-1}$, where τ_x is a translation by x).

If a set is invariant under $\tau G \tau^{-1}$ for some $\tau \in \mathrm{Tr}$, we shall say that it is *invariant under* G *up to a translation*, or simply (if it does not lead to confusion) *G-invariant*.

For any group $G \subset O(n)$, any convex body A, and a point $x \in A$, let

$$A_{x,G} := \bigcap_{g \in G} \tau_x g \tau_x^{-1}(A). \tag{12.22}$$

(Compare (12.18).)

12.6.6. PROPOSITION. *For every group* $G \subset O(n)$ *and any convex bodies* A, C *in* $\mathbb{R}^n$, *the following conditions are equivalent:*

(i) C *is a maximal (with respect to the volume) G-invariant convex body contained in* A;

(ii) *there exists* $p \in A$ *such that* $C = A_{p,G}$, *and the function* $\psi_G : A \to \mathbb{R}$ *defined by*

$$\psi_G(x) := V_n(A_{x,G}) \tag{12.23}$$

attains its upper bound at p.

(See Exercise 12.15.)

Every convex body C satisfying condition (i) (and so also (ii)) of 12.6.6 will be called a *G-kernel of A*, and every point p for which $A_{p,G}$ is a G-kernel of A will be called a *G-pseudocenter of A*.

The set of G-pseudocenters of A will be denoted by $P_G(A)$.

12.6.7. THEOREM. *For any nontrivial subgroup G of $O(n)$, the following conditions are equivalent:*
(i) *every $A \in \mathcal{K}_0^n$ has a unique G-kernel;*
(ii) $G = \{\mathrm{id}_{\mathbb{R}^n}, \sigma_0\}$.

For the implication (ii) $\Rightarrow$ (i) see Theorem 12.6.1. Proof of the converse implication is omitted.[2]

The following problem is open.

12.6.8. PROBLEM. Characterize the class of convex bodies that have a unique G-kernel for every subgroup G of $O(n)$.

The following theorem gives a partial solution of this problem (see [52] Theorem 3.9).

12.6.9. THEOREM. *If A is strictly convex, then for every nontrivial subgroup G of $O(n)$ there is a unique G-kernel of A.*

Proof. Suppose A has at least two G-kernels, C_0 and C_1. By 12.6.6, they are of the form $C_i = A_{p_i,G}$ for some $p_0, p_1 \in P_G(A)$.

Let $p_t := (1-t)p_0 + tp_1$ for $t \in [0;1]$ and $C_t := (1-t)C_0 + tC_1$. Then

$$A_{p_t,G} \supset C_t \qquad (12.24)$$

(see [52]; compare Exercise 12.17).

From (12.24) and the Brunn–Minkowski inequality (Theorem 5.2.2) it follows that

$$\psi_G(p_t) \geq \left((1-t)\psi_G(p_0)^{\frac{1}{n}} + t\psi_G(p_1)^{\frac{1}{n}}\right)^n.$$

Since $\psi_G(p_0) = \psi_G(p_1)$, we obtain

$$\psi_G(p_t) \geq \psi_G(p_i) \qquad (12.25)$$

for $i = 1, 2$. But the function ψ_G attains its upper bound at p_i, whence in (12.25) we have equality. This corresponds to the equality in the Brunn–Minkowski inequality for C_0 and C_1. Since these two sets have equal volumes, it follows that $C_1 = C_0 + v$ for some $v \neq 0$.

Since A is strictly convex, for every $c \in C_0$ the relative interior of $\Delta(c, c+v)$ is contained in $\mathrm{int}A$, and thus

$$C_{\frac{1}{2}} \subset \mathrm{int}A.$$

Let

---

[2] In the proof of Theorem 3.8 in [52], in the last formula and in (3.4), the sign "=" should be replaced by "≤".

$$\varepsilon := \mathrm{dist}\big(C_{\frac{1}{2}}, \mathrm{bd}\,A\big) \text{ and } C := C_{\frac{1}{2}} + \varepsilon B^n.$$

Then C is a G-invariant subset of A and $V_n(C) > V_n\big(C_{\frac{1}{2}}\big) = V_n(C_i)$ (as $\varepsilon > 0$), contrary to the assumption. $\square$

In the case of the group $G = \langle \sigma_0 \rangle$ (i.e., the group generated by the central symmetry σ_0), for the (unique) G-kernel of A there exists a unique G-pseudocenter, $p(A)$.

For an arbitrary G the situation may be much more complicated:

12.6.10. EXAMPLE. Let $G = \langle \sigma_L \rangle$ for a line L being an axis of symmetry of a convex body A. Then A is a unique G-kernel of itself, while every point of $L \cap A$ is a G-pseudocenter of A.

Hence, the following problem arises:

12.6.11. PROBLEM. When is $P_G(A)$ a singleton?

A partial answer to the question 12.6.11 is given by Theorem 12.6.13, which follows from Theorem 12.6.9 combined with the following lemma (compare 3.10 in [52]):

12.6.12. LEMMA. *Let $A \in \mathcal{K}_0^n$ and let G be an arbitrary subgroup of $O(n)$. If 0 is the unique fixed point of the group G, then for every $x, y \in P_G(A)$,*

$$A_{x,G} = A_{y,G} \Longrightarrow x = y.$$

Proof. Suppose $x \neq y$. The set $P_G(A)$ is convex (Exercise 12.16), whence its intersection with $A_{x,G}$ is convex as well, and thus $\Delta(x, y)$ is contained in this intersection. Therefore, we may assume that $x \neq -y$, because otherwise, x, y can be replaced by points of the segment $\Delta(x, y)$ that satisfy this condition.

By the assumption on G, there exists $g \in G$ with

$$g(x + y) \neq x + y. \tag{12.26}$$

Since for every $p \in A$ the set $A_{p,G}$ is invariant under the map $g_p := \tau_p g \tau_p^{-1}$, it follows that $A_{x,G}$ (which coincides with $A_{y,G}$) is invariant under both g_x and g_y; thus it is also invariant under $f := g_x g_y^{-1}$.

However, let us observe that for every $z \in \mathbf{R}^n$,

$$f(z) = z + (x + y) - g(x + y),$$

i.e., f is a translation by the vector $v = (x+y) - g(x+y)$. Since $A_{x,G}$ is compact, it follows that $v = 0$. Hence $g(x + y) = x + y$, contrary to (12.26). $\square$

12.6.13. THEOREM. *Let G be a subgroup of $O(n)$. If 0 is the only fixed point of the group G, then every strictly convex body has a unique G-pseudocenter.*

If $P_G(A)$ is a singleton, then the unique G-pseudocenter of A will be denoted by $p_G(A)$.

Hence we arrive at the following.

12.6.14. COROLLARY. *Let G be a subgroup of $O(n)$. If 0 is the only fixed point of the group G, then the function p_G restricted to the family of strictly convex bodies in R^n is a selector for this family.*

Let us mention that the G-pseudocenter is always equivariant under translations (compare Proposition 2.1 in [53]), but generally, it is not equivariant with respect to an arbitrary isometry of R^n (see Example 2.2 in [53]). For this reason, Rolf Schneider suggested that one consider a modified notion, which was studied in [53].

We end this section with the following theorem, which says that if G is a subgroup of $O(n)$ with only one fixed point 0, then for most convex bodies in R^n there is a unique G-pseudocenter.

To be more precise, let us recall that a subset of a topological space is *meager* if it is a countable union of nowhere dense sets, while a *residual* set is the complement of a meager set (see [64], p. 119).

12.6.15. THEOREM. *Let $\mathcal{X}$ be a subclass of $\mathcal{K}_0^n$ defined as follows:*
$A \in \mathcal{X}$ if and only if $\operatorname{card} P_G(A) = 1$ for every subgroup G of $O(n)$ with the unique fixed point $\{0\}$.
Then $\mathcal{X}$ is residual in $\mathcal{K}^n$.

Proof. By Theorem 12.6.13, all strictly convex bodies in R^n belong to $\mathcal{X}$. Since, in view of Theorem 2.6.1 in [64], the class of strictly convex bodies is residual in $\mathcal{K}^n$, it follows that $\mathcal{X}$ is residual.   □

12.7 G-quasi-centers. Chebyshev point

In [17], Fáry and Rédei asked a question that can be formulated as follows: Does the statement 12.6.1 remain true if the centrally symmetric convex body $C \subset A$ with maximal volume is replaced by a centrally symmetric convex body containing A with minimal volume? They noticed that the answer is negative (Example 12.7.2).

Any centrally symmetric convex body C containing A with minimal volume will be called a *centrally symmetric hull of A*, and the symmetry center of C will be called a *quasi-center of A*.

Let $Q(A)$ be the set of quasi-centers of A.

12.7.1. PROPOSITION. (i) *For every $A \in \mathcal{K}^n$ and $x \in A$, the set $\operatorname{conv}(A \cup \sigma_x(A))$ is the smallest convex body symmetric at x, containing A.*
(ii) *The function $x \mapsto V_n(\operatorname{conv}(A \cup \sigma_x(A)))$ is continuous.*

Proof. (i) is obvious; (ii) follows from 3.2.11 combined with 7.3.3.   □

12.7.2. EXAMPLE. Let A be a triangle, $A = \Delta(a_0, a_1, a_2) \subset R^2$. There exists infinitely many centrally symmetric convex bodies with minimal area containing A. The set $Q(A)$ is the triangle with vertices $b_k := \frac{1}{2}(a_i + a_j)$ for distinct $i, j, k \in \{0, 1, 2\}$.

Indeed, let $X := \Delta(b_0, b_1, b_2)$ and $x \in A$ (Figure 12.4).

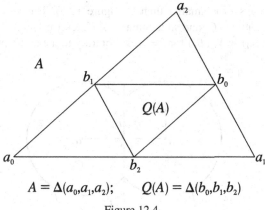

$$A = \Delta(a_0, a_1, a_2); \qquad Q(A) = \Delta(b_0, b_1, b_2)$$

Figure 12.4.

If $x \in \text{int}X$, then $\text{conv}(A \cup \sigma_x(A))$ is the hexagon with vertices $a_0, \sigma_x(a_0), a_1,$ $\sigma_x(a_1), a_2, \sigma_x(a_2)$. Its area equals $2V_2(A)$. By 12.7.1 (ii), the same holds for $x \in \text{bd}X$.

If $x \notin X$, then $x \in \Delta(a_i, b_j, b_k)$ for distinct i, j, k, and the set $\text{conv}(A \cup \sigma_x(A))$ is a parallelogram with area $\|a_i - a_k\| \cdot 2\varrho(x, \Delta(a_j, a_k)) > 2V_2(A)$.

Hence $X = Q(A)$.   □

One may suspect that triangles (or generally, simplices) are the only convex bodies with more than one quasi-center. However, it turns out that the situation is different:

12.7.3. EXAMPLE. (T. Żukowski) For every odd $n \geq 3$ there exists a convex n-gon in $\mathbb{R}^2$ with an infinite set of quasi-centers; for instance, any regular n-gon has this property. The proof is left to the reader (Exercise 12.19).

Let us consider the following more general problem, in some sense dual to 12.6.5.

12.7.4. PROBLEM. Let G be a subgroup of $O(n)$. Is it true that for every $A \in \mathcal{K}_0^n$ there exists a unique G-invariant (up to a translation) convex body with minimal volume, containing A?

Examples 12.7.2 and 12.7.3 show that for $G = \langle \sigma_0 \rangle$ the answer to this question is negative.

The following theorem gives the positive answer for $G = O(n)$.

12.7.5. THEOREM. *For every $A \in \mathcal{K}^n$ there is a unique ball with minimal volume containing A. Its center belongs to A.*

Proof. Suppose that there are two such minimal balls, B_1 and B_2,

$$B_i = B(a_i, r) \text{ for } i = 1, 2.$$

Let $C := B_1 \cap B_2$ and let H be the bisectrix hyperplane of the segment $\Delta(a_1, a_2)$. The set $H \cap C$ is an $(n-1)$-dimensional ball; its center c is the symmetry center of C, and its radius r_0 is smaller than r (Figure 12.5). The n-dimensional ball $B_0 := B(c, r_0)$ contains C, and so it contains A (Exercise 12.18). Since $r_0 < r$, it follows that $V_n(B_0) < V_n(B_i)$ for $i = 1, 2$, contrary to the assumption.

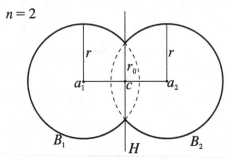

Figure 12.5.

Further, for any $x \in \mathbb{R}^n$, let $R_A(x)$ be the radius of the minimal ball with center x, containing A (compare (12.8)). Observe that if $x \notin A$, then there is a $y \in \mathbb{R}^n$ such that

$$R_A(x) > R_A(y). \tag{12.27}$$

Indeed, for some α, a support hyperplane H of $B(x, \alpha)$ separates x from A; if $y \in H \cap B(x, \alpha)$, then for every $z \in A$,

$$\|x - z\| > \|y - z\|,$$

which implies (12.27). It is easy to check that by the continuity of R_A ([46]), this completes the proof. □

The center of the ball with minimal radius (i.e., of the minimal volume) containing a compact convex set A is called the *Chebyshev point of A*. We denote it by $\check{c}(A)$.

As a direct consequence of 12.7.5, we obtain

12.7.6. COROLLARY. *The map* $\check{c} : \mathcal{K}^n \to \mathbb{R}^n$ *is a selector for* $\mathcal{K}^n$.

It is easy to prove that the map $\check{c}$ is equivariant under similarities (Exercise 12.20). Notice that $\{\check{c}(A)\}$ is the unique singleton nearest to A in the sense of the Hausdorff metric (compare Exercise 12.21).[3]

Generally, the following notion of G-quasi-center is a natural counterpart of that of G-pseudocenter:

---

[3] For related results see [6].

12.7.7. DEFINITION. Let G be a subgroup of $O(n)$. A point $x \in A$ is a *G-quasi-center of a convex body* A if there exists a convex body with minimal volume containing A that is invariant under the group $\tau_x G \tau_x^{-1}$.

Let $Q_G(A)$ be the set of *G*-quasi-centers of A. If $Q_G(A)$ is a singleton, let $q_G(A)$ be the unique *G*-quasi-center. Thus in particular,

$$\check{c} = q_{O(n)}.$$

The following statement is a generalization of 12.7.1.

12.7.8. PROPOSITION. *Let G be a subgroup of $O(n)$.*

(i) *For every $A \in \mathcal{K}^n$ and $x \in A$, the set* $\operatorname{conv} \bigcup_{g \in G} g(A - x) + x$ *is the smallest convex body invariant with respect to $\tau_x G \tau_x^{-1}$ containing A.*

(ii) *The function $x \mapsto V_n \left(\operatorname{conv} \bigcup_{g \in G} g(A - x) + x \right)$ is continuous.*

Proof is analogous to that of 12.7.1.   □

As a consequence of Proposition 12.7.8, we obtain

12.7.9. THEOREM. *For every subgroup G of $O(n)$ and every $A \in \mathcal{K}_0^n$,*

$$Q_G(A) \neq \emptyset.$$

The following problem is open:

12.7.10. PROBLEM. Is it true that there is no proper subgroup G of $O(n)$ such that $Q_G(A)$ is a singleton for every $A \in \mathcal{K}_0^n$?

We shall return to selectors for convex bodies in Chapters 13 and 16. Section 13.5 deals with the Santaló point map, which is defined in terms of polarity. In Section 16.1 we consider radial center maps, which are restrictions to $\mathcal{K}_0^n$ of some multiselectors for star bodies.

13
Polarity

Among transformations of $\mathcal{K}^n$ in $\mathcal{K}^n$ (compare Chapter 4) a particular role is played by polarity. For this reason we devote a separate chapter to this operation.

Let us first briefly recall the notion of a polar hyperplane. In fact, this notion belongs to the theory of algebraic sets in n-dimensional projective space. In that theory the polar hyperplane of a point with respect to an algebraic manifold of degree 2 is defined ([11], [50]). However, the only manifolds we need are spheres, hence we shall deal with $\mathbf{R}^n$, which can be treated as the set of proper points in projective space.

13.1 Polar hyperplane of a point with respect to the unit sphere

We begin with the equation of the hyperplane tangent to a sphere S at a, i.e., the support hyperplane of the ball B with $\mathrm{bd}\,B = S$ at that point.

13.1.1. PROPOSITION. *Let S be the sphere in $\mathbf{R}^n$ with center x_0 and radius r,*

$$S = r \cdot S^{n-1} + x_0,$$

and let $a \in S$. Then the hyperplane tangent to S at a is described by the equation

$$(a - x_0) \circ (x - x_0) = r^2. \tag{13.1}$$

This well-known fact can be proved by means of algebraic methods (compare [50]) or methods of differential geometry, but it can also be proved as follows.

Proof. Since S is symmetric with respect to every line passing through x_0, it easily follows that the hyperplane tangent at a (as the union of lines that have only the point a in common with S) is orthogonal to the vector $a - x_0$. Thus it is described by the equation $(a - x_0) \circ (x - a) = 0$, which is equivalent to $(a - x_0) \circ (x - x_0) = r^2$ because $\|a - x_0\| = r$.   □

13.1.2. DEFINITION. Let $S := \mathrm{bd}\, B(x_0, r)$ for some $r > 0$ and $x_0 \in R^n$. For every $a \in R^n \setminus \{x_0\}$,

$$B_S(a) := \{x \in R^n \mid (a - x_0) \circ (x - x_0) = r^2\}.$$

The hyperplane $B_S(a)$ is called the *polar hyperplane of a with respect to S*.

Directly from Definition 13.1.2 it follows that

$$b \in B_S(a) \implies a \in B_S(b). \tag{13.2}$$

This simple property can be used to construct the polar hyperplane of any point $a \in R^n \setminus \{x_0\}$ with respect to S (Figure 13.1).

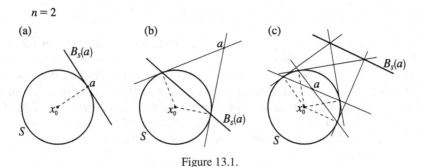

Figure 13.1.

Let $C_S(a)$ be the *cone with vertex a tangent to S*, i.e., the union of all lines tangent to S and passing through a.

13.1.3. THEOREM. *Let $S = \mathrm{bd}\, B(x_0, r)$.*
(i) *If $a \in S$, then $B_S(a)$ is the hyperplane tangent to S at a.*
(ii) *If $a \in R^n \setminus B(x_0, r)$, then*

$$B_S(a) = \mathrm{aff}(C_S(a) \cap S).$$

(iii) *If $a \in \mathrm{int}\, B(x_0, r) \setminus \{x_0\}$, hyperplanes $H_1, \dots, H_n$ intersect at a, and $H_i \cap S = C_S(b_i) \cap S$ for $i = 1, \dots, n$, then*

$$B_S(a) = \mathrm{aff}(b_1, \dots, b_n).$$

Proof. (i) follows directly from 13.1.1 combined with 13.1.2.
(ii): Let us observe that

$$C_S(a) \cap S = B_S(a) \cap S. \tag{13.3}$$

Indeed,

$$x \in C_S(a) \cap S \Longleftrightarrow$$
$$(x - x_0) \circ (a - x) = 0 \ \text{ and } \ (x - x_0) \circ (x - x_0) = r^2 \Longleftrightarrow$$
$$(x - x_0) \circ (a - x_0) = r^2 \ \text{ and } \ (x - x_0) \circ (x - x_0) = r^2 \Longleftrightarrow x \in B_S(a) \cap S.$$

By (13.3),

$$\text{aff}(C_S(a) \cap S) \subset B_S(a). \tag{13.4}$$

Since $\text{dist}(x_0, B_S(a)) = \frac{r^2}{\|a - x_0\|} < r$, it follows that the set $B_S(a) \cap S$ (and thus also $C_S(a) \cap S$) is a sphere of dimension $n - 2$, whence it generates a hyperplane. Thus in (13.4) equality holds.

(iii): Condition (13.3) is satisfied for every point outside the ball $B(x_0, r)$, in particular for b_i:

$$C_S(b_i) \cap S = B_S(b_i) \cap S;$$

hence $H_i = B_S(b_i)$ for $i = 1, \ldots, n$. Therefore, $a \in B_S(b_i)$, and so by (13.2), $b_i \in B_S(a)$ for $i = 1, \ldots, n$. Thus

$$\text{aff}(b_1, \ldots, b_n) \subset B_S(a).$$

It remains to show that $\dim \text{aff}(b_1, \ldots, b_n) = n - 1$, which is equivalent to

$$\det(b_1 - x_0, \ldots, b_n - x_0) \neq 0. \tag{13.5}$$

Since the system of equations

$$(x - x_0) \circ (b_i - x_0) = 0 \ \text{ for } \ i = 1, \ldots, n,$$

which describes $\bigcap_{i=1}^n H_i$, has the matrix $((b_1 - x_0)^T, \ldots, (b_n - x_0)^T)$, condition (13.5) follows from the assumption $\bigcap_{i=1}^n H_i = \{a\}$. $\square$

13.2 Polarity for arbitrary subsets of $\mathbf{R}^n$

For arbitrary nonempty subsets of $\mathbf{R}^n$, polarity is defined as follows.

13.2.1. DEFINITION. Let $A \subset \mathbf{R}^n$, $A \neq \emptyset$. The set A^* defined by the formula

$$A^* := \{x \in \mathbf{R}^n \mid \forall a \in A \ x \circ a \leq 1\}$$

is *polar* to A; the function $A \mapsto A^*$ is the *polarity*.

The relationship between the notion of polarity and that of polar hyperplane is evident (see 13.2.2). For every $a \in \mathbf{R}^n$, let

$$B(a) := B_{S^{n-1}}(a)$$

and

$$E(a) := \{x \in \mathbf{R}^n \mid x \circ a \leq 1\}. \tag{13.6}$$

Then $E(0) = \mathbf{R}^n$, and for $a \neq 0$, the set $E(a)$ is the closed half-space with $\text{bd} E(a) = B(a)$ and $0 \in E(a)$.

13.2.2. PROPOSITION. *For every nonempty subset A of* R^n,

$$A^* = \bigcap \{E(a) \mid a \in A\} = \bigcap \{E(a) \mid a \in A \setminus \{0\}\}.$$

13.2.3. EXAMPLES. (a) $\{a\}^* = E(a)$;
(b) $(B^n)^* = B^n = (S^{n-1})^*$;
(c) $(R^n)^* = \{0\}$.
(Compare Exercise 13.2.)

The following three theorems describe basic properties of polarity.

13.2.4. THEOREM. *For all nonempty subsets* A_1, A_2, *and* A *of* R^n,
(i) $A_1 \subset A_2 \Longrightarrow A_1^* \supset A_2^*$;
(ii) $(A_1 \cup A_2)^* \subset A_1^* \cap A_2^*$;
(iii) $A \subset A^{**}$.
Proof. (i) follows from 13.2.2; since

$$A_i \subset A_1 \cup A_2 \text{ for } i = 1, 2,$$

(i) implies (ii).
(iii): $y \in A \Longrightarrow \forall x \in A^* \ x \circ y \leq 1 \Longrightarrow y \in A^{**}$.   □

13.2.5. THEOREM. *Let* $\emptyset \neq A \subset R^n$.
(i) *For every* $t > 0$

$$(tA)^* = \frac{1}{t} A^*.$$

(ii) *For every linear isometry* $f : R^n \to R^n$,

$$f(A^*) = (f(A))^*.$$

Proof. (i): Since

$$(tx) \circ a = t(x \circ a) = x \circ (ta),$$

it follows that

$$(tA)^* = \{x \in R^n \mid \forall a \in A \ (tx) \circ a \leq 1\}$$
$$= \frac{1}{t} \{tx \in R^n \mid \forall a \in A \ (tx) \circ a \leq 1\} = \frac{1}{t} A^*.$$

(ii): In turn, for every $f \in O(n)$,

$$f(x) \circ f(a) = x \circ a;$$

thus

$$(f(A))^* = \{f(x) \in R^n \mid \forall a \in A \ f(x) \circ f(a) \leq 1\}$$
$$= f\{x \in R^n \mid \forall a \in A \ x \circ a \leq 1\} = f(A^*).$$ □

13.2.6. THEOREM. *Let A be a nonempty subset of* R^n.
(i) *If A is bounded, then* $0 \in \text{int} A^*$.
(ii) *If* $0 \in \text{int} A$, *then* A^* *is bounded.*

Proof. (i): Let $A \subset r B^n$ for some $r > 0$. Then by 13.2.4 (i), 13.2.5.(i), and 13.2.3 (b),

$$A^* \supset \frac{1}{r}(B^n)^* = \frac{1}{r}B^n,$$

whence $0 \in \text{int} A^*$.

(ii): Let $0 \in \text{int} A$. Then $r B^n \subset A$ for some $r > 0$; thus by 13.2.4 (i), 13.2.3 (b), and 13.2.5 (i),

$$A^* \subset (r B^n)^* = \frac{1}{r}B^n,$$

whence A^* is bounded. $\square$

13.3 Polarity for convex bodies

We shall now restrict our considerations to the family $\mathcal{K}_{00}^n$ of convex bodies with 0 in the interior

$$\mathcal{K}_{00}^n := \{A \in \mathcal{K}^n \mid 0 \in \text{int} A\}.$$

From 13.2.2 it follows that for any nonempty $A \subset R^n$ the set A^* is closed and convex. Hence, 13.2.6 yields the following.

13.3.1. THEOREM. $A \in \mathcal{K}_{00}^n \implies A^* \in \mathcal{K}_{00}^n$.

We shall now prove that polarity is idempotent:

13.3.2. THEOREM. *If $A \in \mathcal{K}_0^n$ and $0 \in A$, then*

$$A^{**} = A.$$

Proof. In view of 13.2.4 (iii), it suffices to prove that

$$A^{**} \subset A. \tag{13.7}$$

Let $x \in R^n \setminus A$. Since A is compact, there exists a hyperplane H such that A and x are in different open half-spaces determined by H. Since $0 \in A$ and

$$H = \{y \in R^n \mid y \circ v = 1\}$$

for some $v \neq 0$, it follows that

$$A \subset \{y \in R^n \mid y \circ v < 1\} \tag{13.8}$$

and

$$x \circ v > 1. \tag{13.9}$$

By (13.8) we infer that $v \in A^*$, whence by (13.9), $x \notin A^{**}$. This completes the proof of (13.7). $\square$

As we shall see, the operation $*$ restricted to $\mathcal{K}_{00}^n$ is continuous with respect to the Hausdorff metric.[1] We need the following lemma.

13.3.3. LEMMA. *For every $A \in \mathcal{K}_{00}^n$ and $u \in S^{n-1}$,*

$$h(A^*, u) = \sup_{v \in S^{n-1}} \frac{u \circ v}{h(A, v)}.$$

Proof. Since $h(A, x) = \sup_{a \in A} a \circ x$, it follows that

$$x \in A^* \iff \forall a \in A \ \ x \circ a \le 1 \iff h(A, x) \le 1.$$

If $v = \frac{x}{\|x\|}$, then $h(A, v) = \frac{1}{\|x\|} h(A, x)$. Hence for every $u \in S^{n-1}$,

$$h(A^*, u) = \sup\{x \circ u \mid h(A, x) \le 1\} = \sup_{x \ne 0} \frac{x \circ u}{h(A, x)} = \sup_{v \in S^{n-1}} \frac{v \circ u}{h(A, v)}. \quad \square$$

13.3.4. THEOREM. *Polarity restricted to $\mathcal{K}_{00}^n$ is continuous with respect to ϱ_H.*
Proof. Let $A_1, A_2 \in \mathcal{K}_{00}^n$. There exists $r > 0$ such that

$$r B^n \subset A_1 \cap A_2.$$

We are going to prove that

$$\varrho_H((A_1)^*, (A_2)^*) \le \frac{1}{r^2} \varrho_H(A_1, A_2). \tag{13.10}$$

By 3.4.10, there exists $u \in S^{n-1}$ satisfying the condition

$$\varrho_H((A_1)^*, (A_2)^*) = |h((A_1)^*, u) - h((A_2)^*, u)|;$$

without any loss of generality we may assume that

$$\varrho_H((A_1)^*, (A_2)^*) = h((A_1)^*, u) - h((A_2)^*, u).$$

In view of Lemma 13.3.3, there exists $v \in S^{n-1}$ such that

$$h((A_1)^*, u) = \frac{u \circ v}{h(A_1, v)}.$$

Thus

$$\varrho_H((A_1)^*, (A_2)^*) = \frac{u \circ v}{h(A_1, v)} - \sup_{w \in S^{n-1}} \frac{u \circ w}{h(A_2, w)} \le \frac{u \circ v}{h(A_1, v)} - \frac{u \circ v}{h(A_2, v)}$$

$$= (u \circ v) \frac{h(A_2, v) - h(A_1, v)}{h(A_1, v) h(A_2, v)} \le r^{-2} \varrho_H(A_1, A_2),$$

which proves (13.10). $\quad \square$

---

[1] The proof is due to K. Przesławski.

The support function of the convex body polar to A is closely related to the radial function of A (see (7.21)):

13.3.5. THEOREM. *For every* $A \in \mathcal{K}_{00}^n$,

$$\rho_A = \frac{1}{h_{A^*}}.$$

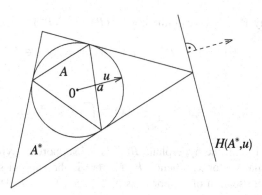

Figure 13.2.

Proof (Figure 13.2). Let $u \in S^{n-1}$ and $a \in (\mathrm{bd}\,A) \cap \mathrm{pos}\,u$. Then $u = \frac{a}{\|a\|}$, and by 13.2.2,

$$H(A^*, u) = B(a) = \{x \in \mathbb{R}^n \mid x \circ a = 1\}.$$

Hence by 3.4.5,

$$h_{A^*}(u) = \mathrm{dist}(0, H(A^*, u)) = \frac{1}{\|a\|} = \frac{1}{\rho_A(u)}. \qquad \square$$

13.4 Combinatorial duality induced by polarity

For convex polytopes, the term "polarity" is often treated as a synonym of "duality" (see Definition 13.4.1). The connection between these two notions is described precisely by Theorem 13.4.3.

13.4.1. DEFINITION. Let $P, P' \in \mathcal{P}^n$.

(a) A bijection $\psi : \mathcal{F}(P) \to \mathcal{F}(P')$ is a *combinatorial duality* for P and P' if ψ reverses inclusion, i.e.,

$$\forall F_0, F_1 \in \mathcal{F}(P) \quad F_0 \subset F_1 \Longrightarrow \psi(F_0) \supset \psi(F_1).$$

(b) The polytopes P and P' are *combinatorially dual* if there exists a combinatorial duality for them.

13.4.2. EXAMPLES. Every two m-gons in $\mathbb{R}^2$ are combinatorially dual, and so are every two n-dimensional simplices in $\mathbb{R}^n$. Any cube and any octahedron in $\mathbb{R}^3$ are combinatorially dual (Exercise 13.4).

Let

$$\mathcal{P}_{00}^n := \mathcal{K}_{00}^n \cap \mathcal{P}^n.$$

13.4.3. THEOREM. *For every $P \in \mathcal{P}_{00}^n$, the polytopes P are P^* are combinatorially dual.*

Proof. For any $P \in \mathcal{P}_{00}^n$ we define $\psi_P : \mathcal{F}(P) \to \mathcal{F}(P^*)$ by the formula

$$\psi_P(F) := \{y \in P^* \mid x \circ y = 1 \text{ for every } x \in F\}. \tag{13.11}$$

Then by 13.1.2,

$$\psi_P(F) = P^* \cap \bigcap_{x \in F} B(x) = \bigcap_{x \in F} P^* \cap B(x).$$

But in view of 13.2.2, the hyperplane $B(x)$ is the supporting hyperplane for P^* with outer normal vector x, whence $P^* \cap B(x)$ is the support set for P^*, and $\psi_P(F)$ is the intersection of support sets of P^* for all $x \in F$. Hence $\psi_P(F) \in \mathcal{F}(P^*)$.

Let us observe that ψ_P is a bijection, because by (13.11) and (13.2),

$$\psi_{P^*}\psi_P = \mathrm{id}_{\mathcal{F}(P)} \tag{13.12}$$

(Exercise 13.5). Directly from (13.11) it follows that ψ_P reverses inclusion. Thus ψ_P is a combinatorial duality. $\square$

By analogy with the notion of face of a convex polytope, a face of an arbitrary convex body A in $\mathbb{R}^n$ can be defined:

13.4.4. DEFINITION. A subset F of a convex body A is called a *face of A* (*proper face of A*) if F is a support set of the body A.

The set of all the faces of A will be denoted by $\mathcal{F}(A)$.

Of course, for polytopes the notion of face in the sense of Definition 13.4.4 coincides with that introduced in Chapter 6.

The notion of combinatorial duality can also be defined generally, for arbitrary convex bodies. We have restricted our consideration to polytopes; generalization is left to the reader (Exercise 13.3). Some remarks can be found in the introduction to [51], where duality is defined in terms of category theory.

13.5 Santaló point

We shall now deal with one more selector for $\mathcal{K}_0^n$, which is closely related to polarity.

Polarity is defined by means of the unit sphere S^{n-1}, and as can be easily seen, it does not commute with translations: generally, $(A - x)^* \neq A^* - x$ (Exercise 13.6). Moreover, generally,

$$V_n((A - x)^*) \neq V_n(A^*) \quad \text{for } x \neq 0 \tag{13.13}$$

(Exercise 13.7).

Thus, the question arises whether for every $A \in \mathcal{K}_0^n$ there exists $x \in \text{int}A$ such that the volume of $(A - x)^*$ is minimal. The following theorem gives an answer to this question (see [64], p. 419).[2]

13.5.1. THEOREM. *For every $A \in \mathcal{K}_0^n$ there is a unique minimizer x_0 of the function $\Psi_A : \text{int}A \to$ R defined by*

$$\Psi_A(x) := V_n((A - x)^*).$$

Idea of proof: the function Ψ_A is strictly convex, whence there is at most one minimizer; further, $\Psi_A(x) \to \infty$ when x approaches bd$A$, which yields the existence of x_0. □

The unique minimizer x_0 of Ψ_A is called the *Santaló point of A*. We denote it by $s_0(A)$.

Let us mention the following interesting relationship between the selectors s_0 and c_{λ_n} ([64], p. 420; [49], Theorem 3.3):

13.5.2. THEOREM. *For every $A \in \mathcal{K}_0^n$,*

$$s_0(A) = x_0 \iff c_{\lambda_n}((A - x_0)^*) = 0.$$

As a direct consequence of 13.5.2, one obtains

13.5.3. COROLLARY. *For every $A \in \mathcal{K}_{00}^n$,*

$$s_0(A) = 0 \iff c_{\lambda_n}(A^*) = 0. \tag{13.14}$$

Finally, let us mention without proof one more theorem on the Santaló point.

13.5.4. THEOREM. *For every $A \in \mathcal{K}_0^n$,*

$$V_n(A) \cdot V_n((A - s_0(A))^*) \leq (\kappa_n)^2. \tag{13.15}$$

Inequality (13.15) is well known as the Blaschke–Santaló inequality ([64] (7.4.24)).

---

[2] A more general version of this theorem can be found in [49].

13.6 Self-duality of the center of the minimal ring

The relationship between the Santaló point and the centroid, described by (13.14) (compare also Exercise 13.8), can be treated as a kind of *duality*. More examples of pairs of selectors dual in this sense are given in [49]. It is a natural question whether there exists a self-dual selector. We shall prove that the answer is positive (Theorem 13.6.2): the center of the minimal ring (see Section 12.5) is self-dual in the above sense. Let us keep the notation used in 12.5.

We need the following lemma (3.4 in [51]).

13.6.1. LEMMA. *Let* $A \in \mathcal{K}_{00}^n$. *If* $A \subset B^n$ *or* $B^n \subset A$, *then*

$$S^{n-1} \cap \mathrm{bd}A = S^{n-1} \cap \mathrm{bd}(A^*). \tag{13.16}$$

Proof. Case 1: $A \subset B^n$. Then $B^n \subset A^*$ by 13.2.3 (b) combined with 13.2.4 (i), whence

$$S^{n-1} \cap \mathrm{bd}A \subset A^*.$$

Since $S^{n-1} \cap \mathrm{bd}A$ is disjoint with $\mathrm{int}(A^*)$, it follows that

$$S^{n-1} \cap \mathrm{bd}A \subset S^{n-1} \cap \mathrm{bd}(A^*).$$

To obtain the converse inclusion, it suffices to prove

$$S^{n-1} \cap \mathrm{bd}(A^*) \subset A. \tag{13.17}$$

Let $a \in S^{n-1} \cap \mathrm{bd}(A^*)$. Then $\mathrm{bd}E(a)$ is a support hyperplane of A^* at a (Exercise 13.9). Since $0 \in E(a) \cap A^*$, it follows that $A^* \subset E(a)$. Hence $a \in E(p)$ for every $p \in A^*$, i.e., $a \in A^{**}$. This proves condition (13.17), because $A^{**} = A$ in view of 13.3.2.

Case 2: $B^n \subset A$. Since in Case 1 the set A was an arbitrary subset of the unit ball, in view of 13.2.4 (i) we may use Case 1 for A replaced by A^*. $\qquad\square$

13.6.2. THEOREM. *For every* $A \in \mathcal{K}_0^n$,

$$c(A) = 0 \implies c(A^*) = 0.$$

Proof. Assume that $c(A) = 0$.
Let

$$\alpha := R_A(0) \quad \text{and} \quad \beta := r_A(0).$$

By definition of r_A and R_A (formulae (12.8) and (12.9)) combined with Theorems 13.2.4 (i) and 13.2.5 (i), we infer that

$$R_{A^*}(0) = \frac{1}{\beta} \quad \text{and} \quad r_{A^*}(0) = \frac{1}{\alpha}. \tag{13.18}$$

By the Bárány Theorem 12.5.6, there exist $p_1, \ldots, p_k \in (\mathrm{bd}A) \cap \alpha S^{n-1}$ and $q_1, \ldots, q_l \in (\mathrm{bd}A) \cap \beta S^{n-1}$ such that

$$\operatorname{conv}\left\{\frac{p_i}{\alpha} \mid i = 1, \ldots, k\right\} \cap \operatorname{conv}\left\{\frac{q_j}{\beta} \mid j = 1, \ldots, \right\} \neq \emptyset. \qquad (13.19)$$

Since $\beta B^n \subset A \subset \alpha B^n$, it follows that

$$\frac{1}{\alpha} A \subset B^n \subset \frac{1}{\beta} A.$$

Thus by Lemma 13.6.1 and Theorem 13.2.5 (i),

$$S^{n-1} \cap \frac{1}{\alpha} \operatorname{bd} A = S^{n-1} \cap \alpha \operatorname{bd}(A^*) \quad \text{and} \quad S^{n-1} \cap \frac{1}{\beta} \operatorname{bd} A = S^{n-1} \cap \beta \operatorname{bd}(A^*).$$

Therefore,

$$\frac{p_i}{\alpha^2} \in \left(\frac{1}{\alpha} S^{n-1}\right) \cap \operatorname{bd}(A^*) \quad \text{for } i = 1, \ldots, k$$

and

$$\frac{q_j}{\beta^2} \in \left(\frac{1}{\beta} S^{n-1}\right) \cap \operatorname{bd}(A^*) \quad \text{for } j = 1, \ldots, l.$$

Now let $p_i' := \frac{p_i}{\alpha^2}$ and $q_j' := \frac{q_j}{\beta^2}$; then $\frac{p_i}{\alpha}$ is the central projection of p_i and p_i' to S^{n-1}, and analogously, $\frac{q_j}{\beta}$ is the central projection of q_j and q_j':

$$\frac{p_i}{\alpha} = \frac{p_i'}{\frac{1}{\alpha}} \quad \text{and} \quad \frac{q_j}{\beta} = \frac{q_j'}{\frac{1}{\beta}}.$$

Consequently, condition (13.19) can be written as

$$\operatorname{conv}\left\{\frac{p_i'}{\frac{1}{\alpha}} \mid i = 1, \ldots, k\right\} \cap \operatorname{conv}\left\{\frac{q_j'}{\frac{1}{\beta}} \mid j = 1, \ldots, l\right\} \neq \emptyset.$$

In view of (13.18), by the Bárány theorem we infer that $c(A^*) = 0$. $\qquad \square$

Theorem 13.6.2 is a particular case of Theorem 3.6 in [51], which will be proved in the next section (see Theorem 13.7.5).

13.7 Metric polarity

Polarity $*$ depends on the sphere S^{n-1}, whence it commutes neither with translations nor with homotheties.

We are now going to define a function $\circledast : \mathcal{K}_0^n \to \mathcal{K}_0^n$, a modified polarity that commutes with all the similarities (Theorem 13.7.4) and thus is called the *metric polarity* (or *metric duality* (see [51]). The idea is to assign to any convex body A some sphere $S(A)$ related to A.

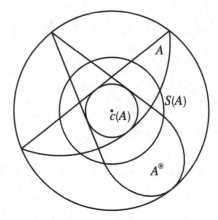

Figure 13.3.

13.7.1. DEFINITION. For every $A \in \mathcal{K}_0^n$,

(i) $\rho(A) := \sqrt{r(A) \cdot R(A)}, \qquad S(A) := \mathrm{bd}\, B(c(A), \rho(A))$;

(ii) $A^{\circledast} := \{x \in \mathbf{R}^n \mid \forall a \in A \ (x - c(A)) \circ (a - c(A)) \leq (\rho(A))^2\}$
(see Figure 13.3).

There is a simple relationship between the operations $*$ and $\circledast$.

13.7.2. PROPOSITION. *Let* $f : \mathbf{R}^n \to \mathbf{R}^n$ *be the similarity defined by*

$$f(x) = \rho(A) \cdot x + c(A). \tag{13.20}$$

Then for every $A \in \mathcal{K}_{00}^n$,

$$f(A^*) = (f(A))^{\circledast}.$$

Proof.

$$f(A^*) = \{f(x) \in \mathbf{R}^n \mid \forall a \in A \ x \circ a \leq 1\}$$
$$= \{y \in \mathbf{R}^n \mid \forall b \in f(A) \ (\rho(A))^{-2}(y - c(A)) \circ (b - c(A)) \leq 1\}$$
$$= (f(A))^{\circledast}. \qquad \square$$

Metric polarity is an involution:

13.7.3. THEOREM. *For every* $A \in \mathcal{K}^n$,

$$A^{\circledast\circledast} = A.$$

Proof. Let f be defined by (13.20) and let $B = f^{-1}(A)$. Then $B \in \mathcal{K}_{00}^n$. Applying twice 13.7.2, we obtain

$$A^{\circledast} = (f(B))^{\circledast} = f(B^*)$$

and

$$A^{\circledast\circledast} = (f(B^*))^{\circledast} = f(B^{**}).$$

Hence by 13.3.2, the assertion follows. $\qquad \square$

Metric polarity commutes with similarities:

13.7.4. THEOREM. *For every similarity $g : R^n \to R^n$ and every $A \in \mathcal{K}_0^n$,*

$$g(A^\circledast) = (g(A))^\circledast.$$

Proof. For any sphere S in R^n and a point a, let $E_S(a)$ be the half-space with boundary $B_S(a)$ such that the center of S belongs to $E_S(A)$. Then in particular,

$$E(a) = E_{S^{n-1}}(a)$$

(compare (13.6)). By 13.7.1 (ii),

$$A^\circledast = \bigcap \{E_{S(A)}(a) \mid a \in A\}. \tag{13.21}$$

Let g be a similarity of R^n. Then for every a and a sphere S,

$$g(E_S(a)) = E_{g(S)}(g(a)) \tag{13.22}$$

and by 13.7.1 (i) and the equivariance of the selector c under similarities (see 12.5.3),

$$g(S(A)) = S(g(A)).$$

Hence by (13.21) and (13.22),

$$g(A^\circledast) = (g(A))^\circledast. \qquad \square$$

The following theorem on metric polarity is a generalization of Theorem 13.6.2.

13.7.5. THEOREM. *For every $A \in \mathcal{K}_0^n$,*

$$c(A^\circledast) = c(A).$$

Proof. Let f be the similarity defined by (13.20). Let us observe that the assertion of Theorem 13.7.2 can be reformulated as follows: for every $A \in \mathcal{K}_0^n$,

$$f^{-1}(A^\circledast) = (f^{-1}(A))^*. \tag{13.23}$$

Since by 12.5.3, $c(f^{-1}(A)) = f^{-1}(c(A)) = 0$, from Theorem 13.6.2 and formula (13.23) it follows that

$$c(f^{-1}(A^\circledast)) = 0.$$

Hence again applying 12.5.3, we obtain

$$f^{-1}(c(A^\circledast)) = 0,$$

whence $c(A^\circledast) = f(0) = c(A)$. $\qquad \square$

Part III

14

Star Sets. Star Bodies

The notion of star body is a natural extension of the notion of convex body. Both classes of sets play an important role in geometric tomography (Gardner's book *Geometric Tomography* [20] was already mentioned in the Introduction). The main task of geometric tomography is, roughly speaking, to examine properties of subsets of R^n by means of properties of their projections and sections.

In the case of projections, it is reasonable to consider only convex bodies because generally, there is no chance to determine a nonconvex set if we know only its projections on hyperplanes: for any subset A of R^n, the projection of A on a hyperplane coincides with the projection of conv$A$.

In the case of intersections, the class of star bodies (or more generally, star sets) is suitable.

14.1 Star sets. Radial function

In the literature, different approaches to the notion of star set can be found. Many authors deal with the notion of a set star-shaped at some point that belongs to this set:

14.1.1. DEFINITION. Let $A \subset R^n$ and $a \in A$. The set A is *star-shaped at a* if

$$\forall x \in A \setminus \{a\} \quad \Delta(a, x) \subset A. \tag{14.1}$$

Gardner and Volčič in [21] introduced the notion of set star-shaped at any point of R^n:

14.1.2. DEFINITION. Let $A \subset R^n$ and $a \in R^n$. The set A is *star-shaped at a* if for every line L passing through a the set $L \cap A$ is connected.

We shall follow Definition 14.1.1. (Gardner in [20] deals with 14.1.2.)

14.1.3. DEFINITION. For every $A \subset R^n$, the *kernel of A* is the subset ker A of all points of A at which A is star-shaped. The set A is called a *star set* if ker $A \neq \emptyset$.

Obviously,

14.1.4. *If A is convex, then* ker $A = A$.

Hence

14.1.5. *Every nonempty convex set is a star set.*

The notion of radial function, which was already defined in Chapter 7 for any convex body A with $0 \in A$ (see (7.21)), can be extended to the class of all bounded sets star-shaped at 0.

14.1.6. DEFINITION. Let A be a bounded subset of R^n and let $0 \in$ ker A. The *radial function* $\rho_A : R^n \setminus \{0\} \to R$ is defined by the formula

$$\rho_A(x) := \sup\{\lambda \geq 0 \mid \lambda x \in A\}.$$

The restriction of ρ_A to S^{n-1} is denoted by the same symbol. Sometimes we write $\rho(A, x)$ instead of $\rho_A(x)$.

Of course, in Definition 14.1.6 the assumption that the set is bounded can be omitted if we allow infinity to be a value of the radial function.

The role of the radial function for star sets is similar to the role of the support function for convex sets. However, there is no simple connection between the radial function of a set and the radial function of its translate. We shall return to this matter in Chapter 16.

Proof of the following simple statement is left to the reader (Exercise 14.4).

14.1.7. PROPOSITION. *If sets $A_1, \ldots, A_k$ are star-shaped at 0, $A = \bigcup_{i=1}^{k} A_i$, and $A_0 = \bigcap_{i=1}^{k} A_i$, then A and A_0 are also star-shaped at 0 and for every $u \in S^{n-1}$,*

$$\rho_A(u) = \max_{1 \leq i \leq k} \rho_{A_i}(u), \quad \rho_{A_0}(u) = \min_{1 \leq i \leq k} \rho_{A_i}(u).$$

As a direct consequence of Definition 14.1.6, we obtain

14.1.8. PROPOSITION. *Let $f \in GL(n)$. For every subset A of R^n star-shaped at 0,*

$$\forall x \neq 0 \quad \rho_{f(A)}(f(x)) = \rho_A(x).$$

For sets that are star-shaped at 0 the notion of *radial sum* is defined as follows:

14.1.9. DEFINITION. (i) For every $x_1, x_2 \in R^n$,

$$x_1 \tilde{+} x_2 := \begin{cases} x_1 + x_2 & \text{if } 0, x_1, x_2 \text{ are collinear;} \\ 0 & \text{otherwise.} \end{cases}$$

(ii) $A_1 \tilde{+} A_2 := \{x_1 \tilde{+} x_2 \mid x_i \in A_i \text{ for } i = 1, 2\}$.

Evidently,

14.1.10. $\rho_{A_1 \tilde{+} A_2} = \rho_{A_1} + \rho_{A_2}$.

It is easy to see (Exercise 14.5) that

14.1.11. $A_1 \tilde{+} A_2 \subset A_1 + A_2$.

14.2 Star bodies

14.2.1. DEFINITION. A nonempty set $A \subset R^n$ is a *body* if A is compact and cl int$A = A$. A body A is called a *star body* if ker $A \neq \emptyset$.

14.2.2. *Every convex body is a star body.*
(Compare Exercise 14.6)

A characterization of convex bodies in terms of radial function is given in [20], Lemma 5.1.4:

14.2.3. THEOREM. *A star body A in R^n with $0 \in$ ker A is convex if and only if for every $u, v \in S^{n-1}$ with $u \neq -v$ and $\rho_A(u), \rho_A(v), \rho_A(u + v) \neq 0$,*

$$\rho_A(u + v)^{-1} \leq \rho_A(u)^{-1} + \rho_A(v)^{-1}.$$

Proof. Let us observe that A is convex if and only if for every $u, v \in S^{n-1}$, $u \neq -v$,

$$x \in (\text{pos}u) \cap \text{bd}A, \ y \in (\text{pos}v) \cap \text{bd}A, \ z \in \text{pos}(u + v) \cap \Delta(x, y)$$
$$\implies z \in A. \tag{14.2}$$

Indeed, condition (14.2) is evidently necessary; the reader can easily prove that it is also sufficient (compare Exercise 14.12).

Hence it suffices to prove that (14.2) is equivalent to the inequality required. Condition (14.2) can be reformulated as follows:

$$z := (1 - t)u\rho_A(u) + tv\rho_A(v) = \|z\|\frac{u + v}{\|u + v\|} \implies \|z\| \leq \|u + v\|\rho_A(u + v).$$

Let $u \neq v$ (for $u = v$ there is nothing to prove). By the linear independence of u, v,

$$(1 - t)\rho_A(u) = \frac{\|z\|}{\|u + v\|} = t\rho_A(v),$$

whence

$$t = \frac{\rho_A(u)}{\rho_A(u) + \rho_A(v)} \quad \text{and} \quad \|z\| = \|u + v\| \frac{\rho_A(u)\rho_A(v)}{\rho_A(u) + \rho_A(v)}.$$

Thus (14.2) is equivalent to

$$\frac{\rho_A(u)\rho_A(v)}{\rho_A(u) + \rho_A(v)} \leq \rho_A(u + v),$$

i.e., is equivalent to

$$\frac{1}{\rho_A(u + v)} \leq \frac{1}{\rho_A(u)} + \frac{1}{\rho_A(v)}. \qquad \square$$

Let us observe that Theorem 7.3.5 and Corollary 7.3.6, concerning convex bodies, remain valid for star bodies with 0 in the kernel. We mention the counterpart of Corollary 7.3.6:

14.2.4. THEOREM. *For every star body A in* R^n *with* $0 \in A$,

$$V_n(A) = \frac{1}{n} \int_{S^{n-1}} \rho_A(u)^n d\sigma(u).$$

In the literature, the name "star body (with respect to 0)" is often used in a more restrictive sense than in Definition 14.2.1. For example, the radial function is assumed to be continuous or continuous on its support, or 0 is assumed to be an internal point of the set (compare [64], [20], [47]).

We shall often restrict our consideration to certain classes of star bodies. The first one, $\mathcal{S}^n$, is defined by means of the support, S_A, of the function $\rho_A|S^{n-1}$; let us recall that

$$S_A := \mathrm{cl}\{u \in S^{n-1} \mid \rho_A(u) \neq 0\}. \qquad (14.3)$$

14.2.5. DEFINITION. $A \in \mathcal{S}^n$ if A is a star body, $0 \in \ker A$, and $\rho_A|S_A$ is continuous.

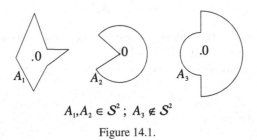

$$A_1, A_2 \in \mathcal{S}^2 \; ; \; A_3 \notin \mathcal{S}^2$$

Figure 14.1.

14.2.6. EXAMPLE. Figure 14.1 presents three star bodies in R^2, only the first two of which, A_1 and A_2, belong to $\mathcal{S}^2$ (Exercise 14.7).

In Chapter 5 we proved the Brunn–Minkowski inequality for the Minkowski sum of convex bodies (Corollary 5.2.2). Lutwak in [43] proved its analogue for the radial sum of star bodies; it is called the *dual Brunn–Minkowski inequality*.

14.2.7. THEOREM ([20], p. 374). *For every $A_1, A_2 \in \mathcal{S}^n$,*

$$V_n(A_1 \widetilde{+} A_2)^{\frac{1}{n}} \leq V_n(A_1)^{\frac{1}{n}} + V_n(A_2)^{\frac{1}{n}}.$$

Equality holds if and only if either $n = 1$ or $n \geq 2$ and $A_2 = \lambda A_1$ for some $\lambda > 0$.[1]

14.3 Radial metric

The family $\mathcal{S}^n$ is not closed in the space of sets star-shaped at 0 with the Hausdorff metric (compare Exercise 14.8). There is another metric, more suitable for this space.

14.3.1. DEFINITION. For any compact subsets A, B of $\mathbb{R}^n$ star-shaped at 0,

$$\delta(A, B) := \sup_{u \in S^{n-1}} |\rho_A(u) - \rho_B(u)|.$$

It is easy to check that the function δ is a metric. It is called the *radial metric*.

14.3.2. PROPOSITION. *For any subsets A_1, A_2 of $\mathbb{R}^n$ star-shaped at 0,*

$$\delta(A_1, A_2) = \inf\{\alpha > 0 \mid A_1 \subset A_2 \widetilde{+} \alpha B^n \text{ and } A_2 \subset A_1 \widetilde{+} \alpha B^n\}. \tag{14.4}$$

Proof. Denote by α_0 the right hand side of (14.4) and let

$$\alpha_i := \inf\{\alpha > 0 \mid A_i \subset A_j \widetilde{+} \alpha B^n\}$$

for $i, j \in \{1, 2\}$, $i \neq j$. It is easy to see that $\alpha_0 = \max\{\alpha_1, \alpha_2\}$, and by 14.1.9,

$$\alpha_i = \inf\{\alpha > 0 \mid \rho_{A_i} \leq \rho_{A_j} + \alpha\}.$$

Thus

$$\begin{aligned}
\alpha_0 &= \inf\{\alpha > 0 \mid |\rho_{A_1} - \rho_{A_2}| \leq \alpha\} \\
&= \sup_{u \in S^{n-1}} |\rho_{A_1}(u) - \rho_{A_2}(u)| = \delta(A_1, A_2). \qquad \square
\end{aligned}$$

As a direct consequence of 14.3.2 combined with 14.1.11, we obtain the following.

---

[1] In [20] (B.28) the case $n = 1$ is omitted.

14.3.3. COROLLARY. *For every compact subsets A_1, A_2 of R^n star-shaped at 0,*

$$\delta(A_1, A_2) \geq \varrho_H(A_1, A_2).$$

We shall prove

14.3.4. THEOREM. *The radial metric is topologically stronger than the Hausdorff metric.*

Proof. From 14.3.3 it follows that radial convergence implies Hausdorff convergence.

It remains to prove that the converse implication does not hold. To this end, let us consider the following example.[2]

Let $L = \mathrm{line}_n$ (where $e_n = (\delta_n^1, \ldots, \delta_n^n)$) and let $a = \frac{1}{4}e_n$. Consider the following sequence $(S_k)_{k\in N}$ of $(n-2)$-dimensional spheres contained in S^{n-1}: the sphere S_k has center $(1 - \frac{1}{k+1})e_n$, and the hyperplane $\mathrm{aff}\,S_k$ is orthogonal to L.

For every $k \in N$, let C_k be the cone with vertex a containing S_k. The set A_k is now defined by

$$A_k := \mathrm{cl}(B^n \setminus \mathrm{conv}\,C_k) \tag{14.5}$$

(Figure 14.2). Then $(A_k)_{k\in N}$ is Hausdorff convergent to B^n, but it is not convergent in the sense of the radial metric, because $\delta(A_k, B^n) = \frac{3}{4}$ for every k.   □

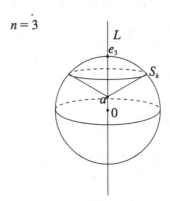

$n = 3$

Figure 14.2.

Let us observe that each of the sets defined by (14.5) is a star body with continuous radial function; i.e., it belongs to $\mathcal{S}^n$. Hence even the restrictions of the Hausdorff metric and the radial metric to $\mathcal{S}^n \times \mathcal{S}^n$ are not topologically equivalent.

---

[2] The two-dimensional version of this example is due to K. Rudnik.

14.4 Star metric

The radial metric has been defined for sets star-shaped at 0. For star sets in the sense of Definition 14.1.3, which are not "fixed at 0," a metric invariant under translations is suitable. Such a metric was introduced in [48]:

14.4.1. DEFINITION. For any compact star sets A, B in $\mathbf{R}^n$, let

$$\vec{\delta}(A, B) := \sup_{x \in \ker A} \inf_{y \in \ker B} \delta(A - x, B - y)$$

and

$$\delta_{st}(A, B) := \max\{\vec{\delta}(A, B), \vec{\delta}(B, A)\} + \varrho_H(\ker A, \ker B).$$

14.4.2. THEOREM. δ_{st} is a metric.

Proof. (a) For every A,

$$\delta_{st}(A, A) = \vec{\delta}(A, A) = \sup_{x \in \ker A} \inf_{y \in \ker A} \delta(A - x, A - y)$$

$$= \sup_{x \in \ker A} \delta(A - x, A - x) = 0.$$

(b) If $\delta_{st}(A, B) = 0$, then $\ker A = \ker B$ and $\vec{\delta}(A, B) = 0 = \vec{\delta}(B, A)$. Thus, since $\ker B$ is compact (compare Exercise 14.1),

$$\forall x \in \ker A \ \exists y \in \ker B \ \delta(A - x, B - y) = 0.$$

But

$$\delta(A - x, B - y) = 0 \Longrightarrow A - x = B - y$$

$$\Longrightarrow \ker A - x = \ker(A - x) = \ker(B - y) = \ker B - y = \ker A - y \Longrightarrow x = y.$$

Hence $A = B$.

(c) Evidently, $\delta_{st}(A, B) = \delta_{st}(B, A)$.

(d) It remains to check the triangle inequality. Let A, B, C be compact star sets. Since for every $x \in \ker A$, $y \in \ker B$, $z \in \ker C$,

$$\delta(A - x, C - z) \leq \delta(A - x, B - y) + \delta(B - y, C - z),$$

it follows that

$$\vec{\delta}(A, C) \leq \vec{\delta}(A, B) + \vec{\delta}(B, C),$$

and similarly,

$$\vec{\delta}(C, A) \leq \vec{\delta}(B, A) + \vec{\delta}(C, B);$$

hence, by the triangle inequality for ϱ_H,

$$\delta_{st}(A, C) \leq \delta_{st}(A, B) + \delta_{st}(B, C). \qquad \square$$

The function δ_{st} is called the *star metric*.

Let us note two simple relationships between the star, the radial, and the Hausdorff metrics (Exercises 14.10 and 14.11):

14.4.3. PROPOSITION. *If* $\ker A = \{a\}$ *and* $\ker B = \{b\}$, *then*

$$\delta_{st}(A, B) = \delta(A - a, B - b) + \|a - b\|.$$

14.4.4. PROPOSITION. *For all compact star sets* A, B *in* $\mathbf{R}^n$, *the following conditions are equivalent:*
(i) $\vec{\delta}(A, B) = 0$;
(ii) $B = A + v$ *for some* $v \in \mathbf{R}^n$;
(iii) $\delta_{st}(A, B) = \varrho_H(A, B)$.

We are going to prove that the star metric is topologically stronger than the Hausdorff metric. We need two lemmas.

14.4.5. LEMMA. *If* A *and* A_k *for* $k \in \mathbf{N}$ *are compact star sets,* $A = \lim_{st} A_k$, *and* $A = \lim_H A_k$, *then for every* $x \in \ker A$ *there exists a sequence* $(x_k)_{k \in \mathbf{N}}$ *such that*
(i) $x_k \in \ker A_k$ *for every* k *and* $\delta(A_k - x_k, A - x) \to 0$;
(ii) $x = \lim x_k$.

Proof. Let $x \in \ker A$. By Definition 14.4.1, there exists a sequence $(x_k)_{k \in \mathbf{N}}$ satisfying condition (i). Since the radial metric is stronger than the Hausdorff metric (Theorem 14.3.4), it follows that

$$\varrho_H(A_k - x_k, A - x) \to 0.$$

But

$$\|x - x_k\| = \varrho_H(A - x, A - x_k) \le \varrho_H(A - x, A_k - x_k) + \varrho_H(A_k - x_k, A - x_k) \to 0,$$

because $\varrho_H(A_k - x_k, A - x_k) = \varrho_H(A_k, A)$. Thus (ii) is satisfied as well. □

14.4.6. LEMMA. *Let* B *and* B_k *for* $k \in \mathbf{N}$ *be compact star sets in* $\mathbf{R}^n$. *If* $B = \lim_H B_k$, *then for every increasing sequence of indices,* $(i_k)_{k \in \mathbf{N}}$, *if* $(\ker B_{i_k})_{k \in \mathbf{N}}$ *is Hausdorff convergent, then*

$$\lim_H \ker B_{i_k} \subset \ker B.$$

Proof. Assume that $(\ker B_{i_k})_{k \in \mathbf{N}}$ is Hausdorff convergent and let $y \in \lim_H \ker B_{i_k}$. Then there exists a sequence $(y_k)_{k \in \mathbf{N}}$ such that $y_k \in \ker B_{i_k}$ for every k and $y = \lim y_k$ (compare Exercise 1.3).

We shall show that $y \in \ker B$. Take arbitrary $z \in B$; since $B = \lim_H B_{i_k}$, it follows that $z = \lim z_k$ for some sequence $(z_k)_{k \in \mathbf{N}}$ with $z_k \in B_{i_k}$. Thus $\Delta(y, z) = \lim_H \Delta(y_k, z_k)$. Since $\Delta(y_k, z_k) \subset B_{i_k}$ for every k, it follows that $\Delta(y, z) \subset B$. □

14.4.7. THEOREM. *The star metric is topologically stronger than the Hausdorff metric.*

Proof. We shall first prove the implication

$$A = \lim_{st} A_k \implies A = \lim_H A_k. \qquad (14.6)$$

Let $A = \lim_{st} A_k$ and $x \in \ker A$. Then $\ker A = \lim_H \ker A_k$ and there exists a sequence $(x_k)_{k \in \mathbb{N}}$ such that $x_k \in \ker A_k$ and $\delta(A_k - x_k, A - x) \to 0$; hence by 14.3.4,

$$A - x = \lim_H (A_k - x_k). \qquad (14.7)$$

In view of the finite compactness of the space $(\mathcal{K}^n, \varrho_H)$ (see 3.2.14), the sequence $(A_k)_{k \in \mathbb{N}}$ has a convergent subsequence A_{i_k} (because of course, it is bounded); let $A' = \lim_H A_{i_k}$. There exists a subsequence $(j_k)_{k \in \mathbb{N}}$ of (i_k) such that $(x_{j_k})_{j \in \mathbb{N}}$ is convergent; let $x' = \lim x_{j_k}$. By Lemma 14.4.6 combined with (14.7), there exists a subsequence (l_k) of (j_k), such that

$$\lim_H \ker(A_{l_k} - x_{l_k}) \subset \ker(A - x);$$

hence $\ker A - x' \subset \ker A - x$, and therefore $x' = x$.

Consequently, by (14.7), $A - x = A' - x$, whence $A' = A$. This completes the proof of (14.6).

To prove that the implication converse to (14.6) does not hold, we can use the example from the proof of Theorem 14.3.4. Indeed, suppose $\lim_{st} A_k = B^n$; then in view of Lemma 14.4.5, for the point $x := e_n$ there exists a sequence $(x_k)_{k \in \mathbb{N}}$ such that $x_k \in \ker A_k$ for every k and $x = \lim x_k$. But as is easy to see, such a sequence does not exist. $\qquad \square$

15
Intersection Bodies

15.1 Dual intrinsic volumes

In Chapter 7 we were concerned with basic functionals for $\mathcal{K}^n$, which are also called intrinsic volumes (compare Definition 7.2.4).

Erwin Lutwak in [43] introduced *dual intrinsic volumes*. To avoid confusion, let us stress that the term "dual" used here has nothing in common with the duality considered in Chapter 13. Dual intrinsic volumes $\tilde{V}_i$ are analogues of the classical intrinsic volumes V_i; they are defined in terms of radial functions. These analogies are presented, e.g., in [64] pp. 385–6.

15.1.1. DEFINITION ([20], (A.55)). For every $A \in \mathcal{S}^n$ and $i \in \mathbb{R}$,

$$\tilde{V}_i(A) := \frac{1}{n} \int_{S^{n-1}} (\rho_A(u))^i \, d\sigma(u).$$

In view of Theorem 14.2.4,

$$\tilde{V}_n(A) = V_n(A). \tag{15.1}$$

By Proposition 14.1.8 and invariance of the spherical measure σ under $O(n)$, we obtain the following.

15.1.2. THEOREM. *All the dual intrinsic volumes are invariant under linear isometries.*

15.2 Projection bodies of convex bodies.
The Shephard problem

The notion of intersection body (for star bodies), which is the main subject of this chapter, is an analogue of the much older notion of projection body (for convex bodies). Hence we begin with projection bodies.

15.2.1. DEFINITION. For any convex body A in R^n, its *projection body*, ΠA, is the convex body defined by the condition

$$\forall u \in S^{n-1} \; h(\Pi A, u) = V_{n-1}(\pi_{u^\perp}(A)). \tag{15.2}$$

(Let us recall that every convex body is uniquely determined by its support function restricted to S^{n-1} (compare Corollary 3.4.6).)

Directly from Definition 15.2.1 it follows that

15.2.2. *Every projection body is symmetric with respect to* 0.

We mention the following two of many well-known characterizations of the class of projection bodies (compare [20], Theorem 4.1.11 and Corollary 4.1.12):

15.2.3. THEOREM. *For every $A \in \mathcal{K}_0^n$ symmetric with respect to* 0 *the following conditions are equivalent:*

(i)   *A is a projection body of a convex body;*
(ii) $A = \lim_H A_i$ *for some sequence of sets A_i, with each A_i being the Minkowski sum of a finite number of segments centered at* 0;
(iii) $A = \lim_H E_i$ *for some sequence of sets E_i, with each E_i being the Minkowski sum of a finite number of rotation ellipsoids with center* 0.

Condition (ii) (without the assumption of symmetry) defines the so-called *zonoids*. In (iii) *rotation ellipsoid* is understood as an ellipsoid with a one-dimensional axis of rotation.

In 1964 Shephard ([65]) asked the following question, usually referred to as the *Shepard problem* (compare [43], Introduction):
Is it true that for any two convex bodies A, B in R^n symmetric with respect to 0, if

$$\forall H \in \mathcal{G}_{n-1}^n \quad V_{n-1}(\pi_H(A)) < V_{n-1}(\pi_H(B)),$$

then

$$V_n(A) < V_n(B) \; ?$$

In 1967, Petty and Schneider independently ([55], [59]) proved that generally the implication is not true, but it is true under the additional assumption that B is a projection body.

Let Π^n be the class of projection bodies.

15.2.4. THEOREM. *Let $A \in \mathcal{K}^n$ and $B \in \Pi^n$. If*

$$\forall H \in \mathcal{G}^n_{n-1} \quad V_{n-1}(\pi_H(A)) \leq V_{n-1}(\pi_H(B)),$$

then $V_n(A) \leq V_n(B)$, and equality holds if and only if B is a translate of A.

A proof of Theorem 15.2.4 can be found in [43].

15.3 Intersection bodies of star bodies. The Busemann–Petty problem

In 1988, Erwin Lutwak introduced the notion of the intersection body of a star body. In the paper cited above, [43], he gave a survey of the theory of projective bodies and in parallel the theory of intersection bodies, and showed analogies between the two concepts.

Lutwak deals with star bodies with the origin in the kernel, with a continuous radial function. The class of such star bodies will be denoted by $\mathcal{S}^n_1$.

15.3.1. DEFINITION. For every $A \in \mathcal{S}^n_1$, let $I A$ be the star body with $0 \in \ker A$ and with radial function defined by the formula

$$\rho_{IA}(u) := V_{n-1}(A \cap u^{\perp}) \quad \text{for every } u \in S^{n-1}.$$

The set $I A$ is called the *intersection body of A* .

The following statement is a consequence of Definition 15.3.1, Theorems 2.5.5 and 2.5.6, and the continuity of V_{n-1}.

15.3.2. PROPOSITION. $A \in \mathcal{S}^n_1 \implies I A \in \mathcal{S}^n_1$.

Evidently,

15.3.3. *Every intersection body is symmetric with respect to 0.*

Goodey and Weil in [23] proved that condition (iii) in Theorem 15.2.3 characterizing projective bodies has its counterpart for intersection bodies. They deal with the class $\mathcal{S}^n_0$ of star bodies that are symmetric with respect to 0 and have continuous radial functions:

$$\mathcal{S}^n_0 := \{A \in \mathcal{S}^n_1 \mid A = -A\}.$$

15.3.4. THEOREM ([23]). *A star body $A \in \mathcal{S}^n_0$ is the intersection body of a star body if and only if A is the radial limit of a sequence of finite radial sums of ellipsoids (with center 0).*

Hence condition (iii) in Theorem 15.2.3 is here replaced by an analogous condition, with $\lim_\delta$ instead of $\lim_H$ and radial addition $\tilde{+}$ instead of the Minkowski addition $+$.

Let us mention that Goodey and Weil used Theorem 15.3.4 to prove that for star bodies, Hausdorff convergence does not imply radial convergence (compare Theorem 14.3.4): in view of Lemma 3 in [23], every $A \in \mathcal{S}_0^n$ is the Hausdorff limit of a sequence of radial sums of ellipsoids; thus if Hausdorff convergence implied radial convergence, then from Theorem 15.3.4 it would follow that every $A \in \mathcal{S}_0^n$ is the intersection body of a star body; however, it is not true (see [20]).

In 1956, Busemann and Petty asked the following question, usually referred to as the *Busemann–Petty problem*:
Is it true that for all convex bodies A, B in R^n centered at 0, if

$$\forall H \in \mathcal{G}_{n-1}^n \quad V_{n-1}(A \cap H) \leq V_{n-1}(B \cap H),$$

then

$$V_n(A) \leq V_n(B) ?$$

Lutwak in [43] deals with the class $\mathcal{S}_1^n$. He proved that the implication does not hold for arbitrary A, $B \in \mathcal{S}_1^n$, but it holds under the additional assumption that A is the intersection body of a star body.

Let $\mathcal{I}^n$ be the class of intersection bodies of members of $\mathcal{S}_1^n$.

15.3.5. THEOREM (Theorem (10.1) in [43]). *Let $A \in \mathcal{I}^n$ and $B \in \mathcal{S}_1^n$. If*

$$\forall u \in S^{n-1} \quad V_{n-1}(A \cap u^\perp) \leq V_{n-1}(B \cap u^\perp),$$

then $V_n(A) \leq V_n(B)$, and equality holds if and only if $A = B$.

The intersection body of a convex body need not be convex:

15.3.6. THEOREM ([20], 8.1.8). *For every $A \in \mathcal{K}_0^n$ there exists a vector v such that $A + v$ has a nonconvex intersection body.*

The following theorem answers the question, when does the function I preserve convexity?

15.3.7. THEOREM ([20], 8.1.11). *If a convex body A in R^n is symmetric with respect to 0, then IA is convex.*

The next three theorems concern the question, when is a convex body (symmetric with respect to 0) the intersection body of a star body?

15.3.8. THEOREM ([20], 8.1.15). *If A is a convex body in R^3, symmetric with respect to 0, with $\rho_A | S^{n-1}$ of the class C^∞, then $A \in \mathcal{I}^n$.*

15.3.9. THEOREM ([20], 8.1.16). *Let C be a rotation cylinder in R^n (that is, C is the direct sum of an $(n-1)$-dimensional ball B with center 0 and a segment with center 0 contained in $(\operatorname{lin} B)^\perp$). Then*

$$C \in \mathcal{I}^n \iff n \leq 3.$$

15.3.10. THEOREM ([20], 8.1.18). *Let Q be an n-cube in R^n with center 0. Then*

$$Q \in \mathcal{I}^n \Longleftrightarrow n \leq 3.$$

The Busemann–Petty problem in its original version (i.e., for convex bodies) has a long and complicated history (see the paper of F. Barthe [3]). Until recently, it has been solved without additional assumptions for $n \neq 4$: the answer to the question is negative for $n \leq 3$ and positive for $n \geq 5$.

G. Zhang in [70], applying results obtained by A. Koldobski, proved the following (Theorem 3 in [22]).

15.3.11. THEOREM. *If A is a convex body in $\mathbf{R}^4$, symmetric with respect to 0, then $A \in \mathcal{I}^4$.*

Earlier, in [69], he found the following characterization of those dimensions for which the answer to the Busemann–Petty question is affirmative (Theorem A in [22]):

15.3.12. THEOREM. *The solution of the Busemann–Petty problem for $\mathbf{R}^n$ is positive if and only if every convex body in $\mathbf{R}^n$ symmetric with respect to 0 is the intersection body of a star body.*

Of course, these two results of Zhang yield the solution for the case $n = 4$.

15.3.13. COROLLARY. *The Busemann–Petty problem has the positive solution for $n \leq 4$ and the negative one for $n \geq 5$.*

15.4 Star duality

Polarity $*$, defined in Chapter 13 for arbitrary nonempty subsets of $\mathbf{R}^n$ (Definition 13.2.1), has interesting properties for the class $\mathcal{K}_{00}^n$ of convex bodies with 0 in the interior: $* : \mathcal{K}_{00}^n \to \mathcal{K}_{00}^n$ is an involution and reverses inclusion.

Metric polarity $\circledast : \mathcal{K}_0^n \to \mathcal{K}_0^n$ is defined in terms of the minimal ring, and thus only for convex bodies (Definition 13.7.1). It is an involution too.

In [51] both functions were extended to functors on some categories with sets of objects $\mathcal{K}_{00}^n$ and $\mathcal{K}_0^n$, respectively. Those functors were called dualities. Here, we put aside the categorical approach, but we preserve the term *duality* for involution.

For the class $\mathcal{S}_+^n$ of star bodies with 0 in the kernel and positive continuous radial function, such a duality $\circ$ was introduced in [47]; it is called the *star duality*.

Let $i : \mathbf{R}^n \setminus \{0\} \to \mathbf{R}^n \setminus \{0\}$ be inversion with respect to S^{n-1}:

$$i(x) := \frac{x}{\|x\|^2}. \tag{15.3}$$

15.4.1. DEFINITION. For every $A \in \mathcal{S}_+^n$,

$$A^\circ := \mathrm{cl}(\mathbf{R}^n \setminus i(A)).$$

15.4.2. PROPOSITION. *For every $A \in \mathcal{S}_+^n$,*

$$\rho_{A^\circ} = \frac{1}{\rho_A}.$$

Proof. Since inversion i is a homeomorphism of $R^n \setminus \{0\}$ onto itself, it follows that bd $i(A) = \mathrm{bd}(R^n \setminus i(A)) = i(\mathrm{bd}A)$. Hence for every $u \in S^{n-1}$, if $a \in \mathrm{pos}\, u \cap \mathrm{bd}A$, then $\rho_A(u) = \|(a)\|$ and $\rho_{A^\circ}(u) = \|i(a)\|$.

In view of (15.3), this completes the proof. □

15.4.3. COROLLARY. *The function $A \mapsto A^\circ$ is an involution that reverses inclusion.*

For convex bodies with 0 in the interior, there is the following relationship between polarity and star duality:

15.4.4. THEOREM. *Let $A \in \mathcal{K}^n_{00}$. Then*

$$\tilde{V}_1(A^\circ) = \frac{\kappa_n}{2} \cdot \bar{b}(A^*).$$

Proof. By Proposition 15.4.2 combined with Theorem 13.3.5, for every $u \in S^{n-1}$,

$$\rho(A^\circ, u) = h(A^*, u). \tag{15.4}$$

Using 15.1.1, 2.2.8, and 3.4.9, we obtain

$$\tilde{V}_1(A^\circ) = \frac{1}{n} \int_{S^{n-1}} \rho(A^\circ, u)d\sigma(u) = \frac{1}{n} \int_{S^{n-1}} h(A^*, u)d\sigma(u)$$
$$= \frac{\kappa_n}{2}\bar{b}(A^*). \qquad\qquad □$$

Generally, for an arbitrary convex body, its polar body is different from the dual star body (Figure 15.1):

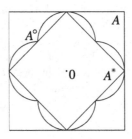

Figure 15.1.

15.4.5. THEOREM. *For every $A \in \mathcal{K}^n_{00}$ the following conditions are equivalent:*

(i) $A^\circ = A^*$;

(ii) $A = \alpha B^n$ *for some $\alpha > 0$.*

Proof. The implication (ii) $\Longrightarrow$ (i) is obvious.

(i) $\Longrightarrow$ (ii): Let $A^\circ = A^*$. Then, by (15.4), $\rho(A^*, u) = h(A^*, u)$ for every $u \in S^{n-1}$; thus for every u,

$$\text{pos}\, u \cap \text{bd}\, A^* \subset H(A^*, u).$$

Therefore, as can be shown, $\text{bd}\, A^*$ is of the class C^1 (Exercise 15.1).

Since for every linear subspace E of dimension 2,

$$E \cap A^\circ = (E \cap A)^\circ,$$

without any loss of generality we may assume that $n = 2$.

Consider a parametrization $r : \text{R} \to \text{bd}\, A^*$ of the boundary of A^*:

$$r(t) := \bar{\rho}(t) \cdot u(t),$$

where $\bar{\rho}(t) = \rho(A^*, u(t))$ and $u(t) = (\cos t, \sin t)$.

Since $r(t) \circ r'(t) = 0$ for every t, it follows that $\bar{\rho} \cdot \bar{\rho}' = 0$. Thus $\bar{\rho}' = 0$, whence $\rho_{A^*} = \text{const}$. Since A^* is a disk, so is A (compare 13.2.3 (b) and 13.2.5 (i)).   □

The question arises whether there are any connections between star duality and the notion of intersection body. Some results related to this question are presented in [47]; they concern connections between star duality and the (slightly more general) notion of the intersection body of order k.

15.4.6. DEFINITION ([20], Note 8.3). For every $A \in \mathcal{S}_1^n$ and $k \in \text{R}$, let $I_k A$ be the set star-shaped at 0 with radial function defined by

$$\rho_{I_k A}(u) := \tilde{V}_k(A \cap u^\perp).$$

The set $I_k A$ is called the *intersection body of order k of A*.

Let $\mathcal{I}_k^n$ be the class of intersection bodies of order k. The following problem is open for arbitrary k as well as for $k = n$:

15.4.7. PROBLEM. Prove or disprove the implication

$$A \in \mathcal{I}_k^n \Longrightarrow A^\circ \in \mathcal{I}_k^n.$$

16
Selectors for Star Bodies

Since properties of star bodies defined in terms of radial functions generally are not invariant under translations, the problem arises what is a proper position of a given star body with respect to 0, or equivalently, how to choose a point that should play the role of the origin. Of course, one has to decide what position of the body with respect to 0 is good.

Some solutions of this problem were suggested in [48].

16.1 Radial centers of a star body

The problem raised above can be formulated as the problem of looking for selectors of the family S^n or some of its subfamilies.

16.1.1. DEFINITION. For any family $\mathcal{F}$ of star bodies in $\mathbf{R}^n$, a function $s : \mathcal{F} \to \mathbf{R}^n$ is a *selector for* $\mathcal{F}$ if $s(A) \in \ker A$ for every $A \in \mathcal{F}$. Similarly, $s : \mathcal{F} \to 2^{\mathbf{R}^n}$ is a *multiselector* if $\emptyset \neq s(A) \subset \ker A$ for every $A \in \mathcal{F}$.

We begin with some multiselectors.

For every star body A and function $\varphi : [0, \infty) \to [0, \infty)$, let us define $\Phi_A : \ker A \to \mathbf{R}$ by the formula

$$\Phi_A(x) := \int_{S^{n-1}} \varphi \rho_{A-x}(u) d\sigma(u). \tag{16.1}$$

We shall consider the subset $M_\varphi(A)$ of $\ker A$ consisting of the points at which Φ_A attains its upper bound, i.e., the set of maximizers of Φ_A. The points of $M_\varphi(A)$ will be called *radial centers of A associated with φ*.

Let us note that if $\varphi(t) = t^\alpha$ for some $\alpha \in (0; 1)$, then $\Phi_A(x)$ is the *dual volume of $A - x$ of order α*:

$$\Phi_A(x) = \tilde{V}_\alpha(A - x)$$

(compare [20] A.55 and [43]). Thus for any φ, the function Φ_A is a *generalized dual volume*.

Let us define a family T^n as follows:

16.1.2. DEFINITION. Let A be a star body in R^n.

$A \in T^n$ if $A \in \mathcal{S}^n$ and there exists $S_0 \subset S^{n-1}$ such that $\sigma(S_0) = 0$ and for every $u \in S^{n-1} \setminus S_0$ the function $\ker A \ni x \mapsto \rho_{A-x}(u) \in R$ is continuous.

It can be proved (see Theorem 2.5 in [48]) that

$$\mathcal{K}_0^n \subset T^n \tag{16.2}$$

and

$$(A \in \mathcal{S}^n \quad \text{and} \quad \ker A \subset \text{int } A) \implies A \in T^n. \tag{16.3}$$

16.1.3. THEOREM. *If $\varphi : [0, \infty) \to R$ is continuous, then for every $A \in T^n$*
(i) *the function Φ_A is continuous;*
(ii) *the set $M_\varphi(A)$ is nonempty and compact.*

Proof. It is easy to see that for every $A \in T^n$ the function Φ_A is continuous. Thus is attains its upper bound because $\ker A$ is compact.   □

In view of 16.1.3, every continuous function $\varphi : [0, \infty) \to R$ induces a multiselector $M_\varphi : T^n \to R^n$.

Generally, M_φ is not a selector. Moreover, it may happen that it does not choose any proper subset of $\ker A$.

16.1.4. EXAMPLE. If $\varphi(t) = t^n$ for every $t \geq 0$, then for every star body A,

$$M_\varphi(A) = \ker A,$$

because Φ_A is constant: for every $x \in \ker A$,

$$\Phi_A(x) = \int_{S^{n-1}} \rho_{A-x}(u)^n d\sigma(u) = n V_n(A)$$

(see Theorem 14.2.4).

As Example 16.1.4 shows, to ensure that the multiselector M_φ is a selector, it is not enough to restrict the class of star bodies, but also additional assumptions on φ have to be admitted.

The following problem is open.

16.1.5. PROBLEM. Is it possible to formulate conditions on φ ensuring that M_φ is a selector for T^n?

The following example shows that generally for star bodies the radial centers associated with the identity are not unique.

16.1.6. EXAMPLE (see [33]). Let A be the subset of $\mathbb{R}^2$ bounded by four arcs of hyperbolas (Figure 16.1):

$$H_1 := \{(x_1, x_2) \mid (x_1 + 1)(x_2 + 1) = 4, \ 0 \le x_i \le 3 \text{ for } i = 1, 2\},$$
$$H_2 := \sigma_{L_2}(H_1), \quad H_3 := \sigma_0(H_1), \quad H_4 := \sigma_{L_1}(H_1),$$

where $L_i = \mathrm{aff}(0, e_i)$ for $i = 1, 2$.

Then ker A is an octagon with one of its vertices $a = (0, \frac{3}{4})$. Suppose A has a unique radial center associated with the identity, $r_{\mathrm{id}}(A)$. Since 0 is the center of symmetry of A, it follows that $r_{\mathrm{id}}(A) = 0$.

However, it can be computed that, approximately, $\Phi_A(0) = 11{,}046$, while $\Phi_A(a) = 21{,}054$; hence $\Phi_A(0) < \Phi_A(a)$, a contradiction.

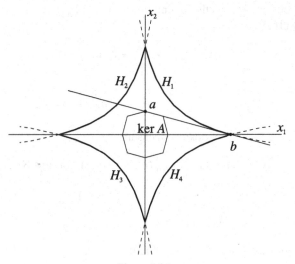

Figure 16.1.

16.2 Radial centers of a convex body

We are going to prove that for $n \ge 2$, under some assumptions on φ every convex body in $\mathbb{R}^n$ has a unique radial center associated with φ (Corollary 16.2.2).

16.2.1. THEOREM ([35]). *Let $n \ge 2$. If $A \in \mathcal{K}_0^n$ and $\varphi : \mathbb{R}_+ \to \mathbb{R}_+$ is concave and strictly increasing, then the function Φ_A defined by (16.1) is strictly concave.*

Proof. Let $x_0, x_1 \in A$, $x_0 \ne x_1$, $t \in (0; 1)$, and let $x := (1 - t)x_0 + tx_1$. Let $A_i := A - x_i$ for $i = 0, 1$. Then

$$A - x = (1 - t)A_0 + tA_1,$$

because A is convex.

Hence by 14.1.11 and by the assumptions on φ,

$$\Phi_A(x) \geq (1 - t)\Phi_A(x_0) + t\Phi_A(x_1). \tag{16.4}$$

It remains to prove that the inequality in (16.4) is sharp.

By the dual Brunn–Minkowski inequality (Theorem 14.2.7), it follows that

$$V_n(A)^{\frac{1}{n}} = ((1 - t)V_n(A_0)^{\frac{1}{n}} + tV_n(A_1)^{\frac{1}{n}}$$
$$\geq V_n((1 - t)A_0 \tilde{+} tA_1)^{\frac{1}{n}}, \tag{16.5}$$

and equality holds if and only if there exists $\lambda > 0$ such that $tA_1 = \lambda(1 - t)A_0$. Comparing the n-volumes, we obtain $\lambda = \frac{t}{1-t}$, which is equivalent to $A - x_0 = A - x_1$; but this contradicts the assumption $x_0 \neq x_1$.

Hence by (16.5),

$$V_n(A - x) > V_n((1 - t)A_0 \tilde{+} tA_1),$$

that is,

$$\int_{S^{n-1}} \rho_{A-x}^n > \int_{S^{n-1}} ((1 - t)\rho_{A_0} + t\rho_{A_1})^n.$$

Thus for some $S \subset S^{n-1}$ of positive measure,

$$\rho_{A-x}(u) > (1 - t)\rho_{A_0}(u) + t\rho_{A_1}(u) \quad \text{for } u \in S,$$

whence for $u \in S$,

$$\varphi\rho_{A-x}(u) > (1 - t)\varphi\rho_{A-x_0}(u) + t\varphi\rho_{A-x_1}(u).$$

Integrating both sides over S^{n-1}, we obtain

$$\Phi_A((1 - t)x_0 + tx_1) > (1 - t)\Phi_A(x_0) + t\Phi_A(x_1),$$

i.e., Φ_A is strictly concave. $\qquad\square$

16.2.2. COROLLARY. *Let $n \geq 2$. If φ is concave and strictly increasing, then $M_\varphi(A)$ is a singleton for every $A \in \mathcal{K}_0^n$.*

Proof. Let $A \in \mathcal{K}_0^n$. Then by (16.2), $A \in \mathcal{T}^n$, whence by Theorem 16.1.3, the function Φ_A is continuous. Thus by 16.2.1, Φ_A has a unique maximizer. $\quad\square$

Let $r_\varphi(A)$ be the unique radial center of a convex body A, induced by φ.

In view of Corollary 16.2.2, if $n \geq 2$ and φ is concave and strictly increasing, then r_φ is a selector for $\mathcal{K}_0^n$.

16.2.3. PROPOSITION. *The selector r_φ is equivariant under isometries.* (Compare Exercise 16.1.)

We shall prove the continuity of the radial centers for convex bodies with respect to the Hausdorff metric.

For every $A \in \mathcal{K}_0^n$, let

$$U(A) := \{u \in S^{n-1} \mid \exists a, b \ \ b - a \in \operatorname{lin} u \text{ and } \Delta(a, b) \subset \operatorname{bd} A\}.$$

We start from the following.

16.2.4. LEMMA ([35]). *Let* $A_k \in \mathcal{K}_0^n$ *and* $x_k \in A_k$ *for every* $k \in \mathbb{N} \cup \{0\}$. *If* $A_0 = \lim_H A_k$ *and* $x_0 = \lim x_k$, *then for every* $u \in S^{n-1} \setminus \bigcup_{k=0}^{\infty} U(A_k)$,

$$\rho_{A - x_0}(u) = \lim \rho_{A_k - x_k}(u).$$

Proof. Without any loss of generality we may assume that $x_k = x_0$ for every k. Indeed, let $A_k' := A_k + x_0 - x_k$; then $\lim_H A_k' = A = \lim_H A_k$ and $\rho_{A_k' - x_0} = \rho_{A_k - x_k}$.

Hence, let $x_0 \in A \cap \bigcap_{k \geq 0} A_k$, $u \in S^{n-1} \setminus \bigcup_{k \geq 0} U(A_k))$, and

$$L^+ := x_0 + \operatorname{pos} u, \quad L^- := x_0 - \operatorname{pos} u.$$

Further, let

$$x_k^+ \in L^+ \cap \operatorname{bd} A_k \quad \text{and} \quad x_k^- \in L^- \cap \operatorname{bd} A_k$$

for every $k \in \mathbb{N} \cup \{0\}$.

It suffices to prove that

$$x_0^+ = \lim x_k^+. \tag{16.6}$$

For every $k \geq 0$ there are the following four possibilities:

$(1)_k \quad x_k^- \neq x_0 \neq x_k^+$,
$(2)_k \quad x_k^- \neq x_0 = x_k^+$,
$(3)_k \quad x_k^- = x_0 \neq x_k^+$,
$(4)_k \quad x_k^- = x_0 = x_k^+$.

Passing to subsequences of $(A_k)_{k \in \mathbb{N}}$, we may assume that $(x_k^+)_{k \in \mathbb{N}}$ is convergent and that exactly one of $(1)_k$–$(4)_k$ is satisfied for all $k \geq 1$, i.e., for all members of the sequence $(A_k)_{k \in \mathbb{N}}$. Consequently, 16 conjunctions $(i)_k \wedge (j)_0$ for $i, j \in \{1, \ldots, 4\}$ are to be considered.

If $L^+ \cap \operatorname{bd} A_0$ is a singleton, i.e., one of the conditions $(i)_k \wedge (1)_0$, $(i)_k \wedge (2)_0$, and $(i)_k \wedge (4)_0$ is satisfied, then (16.6) holds. Thus the 16 conditions are reduced to $(i)_k \wedge (3)_0$ for $i \in \{1, \ldots, 4\}$.

But in view of Theorem 2.5.7, by backward induction on the dimension of the affine subspace, the above conjunction may hold only for $i = 3$, which means that the only possibility is

$$x_k^- = x_0 = x_k^+ \quad \text{and} \quad x_0^- = x_0 = x_0^+,$$

which implies (16.6). $\square$

16.2.5. THEOREM ([35]). *For every $n \geq 2$ and every concave and strictly increasing function φ, the selector $r_\varphi : \mathcal{K}_0^n \to \mathbf{R}^n$ is continuous with respect to the Hausdorff metric.*

Proof. Let $A = \lim_H A_k$ for a sequence $(A_k)_{k \in \mathbf{N}}$ of convex bodies in $\mathbf{R}^n$, and let $x_k = r_\varphi(A_k)$ for every k. We may assume that $(x_k)_{k \in \mathbf{N}}$ is convergent. Let $x = \lim x_k$.

From Lemma 16.2.4 we deduce that

$$\Phi_A(x) = \lim \Phi_{A_k}(x_k). \tag{16.7}$$

We shall prove that $x = r_\varphi(A)$. Suppose, to the contrary, that there exists $y \in A$ such that

$$\Phi_A(y) > \Phi_A(x). \tag{16.8}$$

Since $A = \lim_H A_k$, there exists $(y_k)_{k \in \mathbf{N}}$ such that $y_k \in A_k$ for every k and $\lim y_k = y$. Since $\Phi_{A_k}(x_k) \geq \Phi_{A_k}(y_k)$ for every k, applying again Lemma 16.2.4 to the sequence $(y_k)_{k \in \mathbf{N}}$ we infer from (16.7) that $\Phi_A(x) \geq \Phi_A(y)$, contrary to (16.8). $\qquad\square$

16.3 Extended radial centers of a star body

Observe that in the proof of Theorem 16.1.3 we did not make use of the fact that Φ_A has all of ker A as its domain; the only thing we needed was that the domain was a compact, convex, and nonempty subset of ker A.

Also, in Theorem 16.2.1 and Corollary 16.2.2 we needed only the fact that the domain of Φ_A was a compact, convex, and nonempty subset of a convex body A.

Hence the statements 16.1.3 (ii) and 16.2.2 can be easily generalized as follows:

16.3.1. THEOREM. *If $\varphi : [0, \infty) \to \mathbf{R}$ is concave and strictly increasing, then for every $A \in \mathcal{T}^n$ and every $C \in \mathcal{K}^n$ contained in* ker A, *the set of maximizers of $\Phi_A|C$ is nonempty and compact.*

16.3.2. THEOREM. *Let $n \geq 2$. If $\varphi : [0, \infty) \to \mathbf{R}$ is concave and strictly increasing, $A \in \mathcal{K}_0^n$, $C \in \mathcal{K}^n$, and $C \subset A$, then the function $\Phi_A|C$ has a unique maximizer.*

Let us denote by $r_\varphi(A, C)$ the unique point of C whose existence is ensured by Theorem 16.3.2.

In view of Theorem 16.3.2, the selector r_φ defined for $\mathcal{K}_0^n$ can be extended to a selector for $\mathcal{S}^n$:

16.3.3. THEOREM. *Let $n \geq 2$. If φ is concave and strictly increasing, then the formula*

$$\tilde{r}_\varphi(A) := r_\varphi(\operatorname{conv} A, \ker A) \quad \text{for } A \in \mathcal{S}^n \tag{16.9}$$

defines a selector $\tilde{r}_\varphi$ for the class $\mathcal{S}^n$. It is an extension of r_φ:

$$\forall A \in \mathcal{K}_0^n \; \tilde{r}_\varphi(A) = r_\varphi(A).$$

Proof. In view of Theorem 16.3.2 applied to the convex body conv A and its subset $C := \ker A$, the function $\tilde{r}_\varphi$ is a selector.

If $A \in \mathcal{K}_0^n$, then conv $A = A = \ker A$, whence by (16.9),

$$\tilde{r}_\varphi(A) = r_\varphi(A, A) = r_\varphi(A). \qquad \square$$

16.3.4. PROPOSITION. *The selector $\tilde{r}_\varphi$ is equivariant under isometries.* (Exercise 16.2.)

It is easy to show that $\tilde{r}_\varphi$ is not continuous with respect to the Hausdorff metric (Exercise 16.3). However, as we shall see, it is continuous with respect to the star metric (compare [48]). We need two lemmas (16.3.6 and 16.3.7) and the following result obtained by I. Herburt.

16.3.5. THEOREM ([31]). *The restriction of the function* conv *to the class $\{A \in \mathcal{S}^n \mid 0 \in \ker A \cap \operatorname{int} A\}$ is continuous with respect to the radial metric.*

16.3.6. LEMMA. *Let $A, A_k \in \mathcal{S}^n$, $x_k \in \ker A_k$, $a_k \in \ker A$ for every $k \in \mathbb{N}$, and $\delta(A_k - x_k, A - a_k) \to 0$.*
If $\varphi : [0, \infty) \to \mathbb{R}$ is a Lipschitz function, then

$$\lim \; |\Phi_{A_k}(x_k) - \Phi_A(a_k)| = 0.$$

Proof. Let λ be the Lipschitz constant of φ. Then

$$|\Phi_{A_k}(x_k) - \Phi_A(a_k)| \leq \lambda \int_{S^{n-1}} |\rho_{A_k - x_k}(u) - \rho_{A - a_k}(u)| d\sigma(u)$$

$$\leq \lambda n \kappa_n \delta(A_k - x_k, A - a_k) \to 0. \qquad \square$$

The other lemma concerns the class $\mathcal{T}_0^n$ defined as follows:

$$\mathcal{T}_0^n := \mathcal{K}_0^n \cup \{A \in \mathcal{T}^n \mid \ker A \subset \operatorname{int} A\}. \tag{16.10}$$

Observe that, by (16.2),

$$\mathcal{T}_0^n \subset \mathcal{T}^n. \tag{16.11}$$

16.3.7. LEMMA. *Let $A, A_k \in \mathcal{T}_0^n$ for $k \in \mathbb{N}$, $A = \lim_{st} A_k$, and let $\varphi : [0, \infty) \to [0, \infty)$ be a Lipschitz function.*
If for every k the function $\Phi_{\operatorname{conv} A_k} | \ker A_k$ attains its upper bound at p_k and $\Phi_{\operatorname{conv} A} | \ker A$ attains its upper bound at p, then

$$\lim \Phi_{\operatorname{conv} A_k}(p_k) = \Phi_{\operatorname{conv} A}(p).$$

Proof. Since $A = \lim_{st} A_k$, there exists a sequence $(a_k)_{k \in \mathbb{N}}$ in $\ker A$ such that

$$\delta(A_k - p_k, A - a_k) \to 0, \tag{16.12}$$

whence by Theorem 14.3.4, $\varrho_H(A_k - p_k, A - a_k) \to 0$. By Theorem 14.4.7, $A = \lim_H A_k$; thus

$$\|p_k - a_k\| \to 0.$$

In view of Theorem 16.3.5, from (16.12) it follows that

$$\delta(\mathrm{conv}\,A_k - p_k, \mathrm{conv}\,A - a_k) \to 0.$$

Applying Lemma 16.3.6 to the convex hulls of the sets A_k and A, we infer that

$$\lim |\Phi_{\mathrm{conv}\,A_k}(p_k) - \Phi_{\mathrm{conv}\,A}(a_k)| = 0. \qquad (16.13)$$

Assume (p_k) to be convergent (we may do so because otherwise, we can repeat this reasoning for every convergent subsequence). Let $x = \lim p_k$. Then $x = \lim a_k$. Since

$$|\Phi_{\mathrm{conv}\,A_k}(p_k) - \Phi_{\mathrm{conv}\,A}(x)|$$
$$\leq |\Phi_{\mathrm{conv}\,A_k}(p_k) - \Phi_{\mathrm{conv}\,A}(a_k)| + |\Phi_{\mathrm{conv}\,A}(a_k) - \Phi_{\mathrm{conv}\,A}(x)|,$$

from (16.13) and Theorem 16.1.3 (i) it follows that

$$\Phi_{\mathrm{conv}\,A}(x) = \lim \Phi_{\mathrm{conv}\,A_k}(p_k).$$

Hence

$$\Phi_{\mathrm{conv}\,A}(p) \geq \lim \Phi_{\mathrm{conv}\,A_k}(p_k). \qquad (16.14)$$

On the other hand, there exists a sequence $(x_k) \in \mathbf{P}_{k=1}^{\infty} \ker A_k$ such that $p = \lim x_k$ and $\delta(A_k - x_k, A - p) \to 0$; thus by Theorem 16.3.5,

$$\delta(\mathrm{conv}\,A_k - x_k, \mathrm{conv}\,A - p) \to 0.$$

Applying again Lemma 16.3.6 to convex hulls, we obtain the inequality

$$\Phi_{\mathrm{conv}\,A}(p) \leq \lim \Phi_{\mathrm{conv}\,A_k}(p_k),$$

which, together with (16.14), completes the proof.   □

16.3.8. THEOREM. *Let $n \geq 2$. For every concave, strictly increasing Lipschitz function $\varphi : [0, \infty) \to [0, \infty)$, the selector $\tilde{r}_\varphi | T_0^n$ is continuous with respect to the star metric.*

Proof. Let A, $A_k \in T_0^n$ and $A = \lim_{st} A_k$. Let $p_k = \tilde{r}_\varphi(A_k)$ and $p = \tilde{r}_\varphi(A)$. On the one hand, by Lemma 16.3.7,

$$\Phi_{\mathrm{conv}\,A}(p) = \lim \Phi_{\mathrm{conv}\,A_k}(p_k). \qquad (16.15)$$

On the other hand, there exists a sequence (a_k) in $\ker A$ such that

$$\delta(A_k - p_k, A - a_k) \to 0.$$

As before, we may assume that (a_k) is convergent. Let $x = \lim a_k$. Then $x = \lim p_k$ by 14.4.7 combined with 14.3.4. Since, by 16.1.3 (i) combined with 16.3.5, the function $\Phi_{\text{conv} A}$ is continuous, from Lemma 16.3.6 it follows that

$$\Phi_{\text{conv} A}(x) = \lim \Phi_{\text{conv} A_k}(p_k).$$

Hence by (16.15), we infer that $\Phi_{\text{conv} A}(x) = \Phi_{\text{conv} A}(p)$, whence $x = p$ because p is the unique maximizer. Therefore,

$$p = \lim p_k. \qquad \qquad \square$$

Exercises to Part I

Chapter 1

1.1. Prove that for any metric space (X, ϱ) the function $\varrho(\cdot, A) : X \to \mathbb{R}$ is a weak contraction:

$$|\varrho(x, A) - \varrho(y, A)| \leq \varrho(x, y).$$

1.2. Explain why the proof of Theorem 1.1.11, which concerns $\mathbb{R}^n$, does not work for $\mathcal{K}^n$.

1.3. Prove that for every bounded sequence of nonempty compact subsets $A, A_1, A_2, \ldots$ of a finitely compact metric space (X, ϱ), the condition

$$A = \lim_H A_k$$

is equivalent to the conjunction of the following two conditions:

   (i) for every increasing sequence of indices $(i_k)_{k \in \mathbb{N}}$ and every sequence $(x_{i_k})_{k \in \mathbb{N}} \in \mathbf{P}_{k=1}^{\infty} A_{i_k}$ convergent in (X, ϱ),

$$\lim x_{i_k} \in A;$$

   (ii) for every $x \in A$ there exists a sequence $(x_k)_{k \in \mathbb{N}} \in \mathbf{P}_{k=1}^{\infty} A_k$ such that

$$x = \lim x_k.$$

(Compare [64], note 3, p. 57.)

1.4. Prove that for arbitrary $A, B \in \mathcal{C}(\mathbf{R}^n)$,

$$\varrho_H(A, B) = \sup_{x \in \mathbf{R}^n} |\varrho(x, A) - \varrho(x, B)|.$$

1.5. Prove that every finitely compact metric space is complete.

1.6. Prove or disprove the following statement: If (X, ϱ) is finitely compact, then so is $(\mathcal{C}^n(X), \varrho_H))$.

1.7. Prove that for nonempty subsets A, B of $\mathbf{R}^n$ and $\delta, \varepsilon > 0$,

$$((A)_\delta)_\varepsilon = (A)_{\delta+\varepsilon}.$$

(Compare 1.1.6.)
   Give an example of a metric space (X, ϱ) for which the inclusion $\supset$ generally fails.

Chapter 2

2.1. Prove that if $A_1, A_2, A_1 \cup A_2 \in \mathcal{K}^n$, then

$$(A_1 \cup A_2) + (A_1 \cap A_2) = A_1 + A_2.$$

2.2. Prove that for every $A \in \mathcal{C}^n$ the function $f : \mathbf{R} \to \mathcal{C}^n$ defined by

$$f(t) := t \cdot A$$

is continuous.

2.3. Prove that
   (i) if $A_1, A_2 \in \mathcal{C}^n$ and $A_1 \cap A_2 \neq \emptyset$, then for every $x \in \mathbf{R}^n$,

$$\varrho(x, A_1 \cap A_2) \geq \max_{i=1,2} \varrho(x, A_i);$$

   (ii) if $A_1, A_2, A_1 \cup A_2 \in \mathcal{K}^n$, then for every $x \in \mathbf{R}^n$,

$$\varrho(x, A_1 \cap A_2) = \max_{i=1,2} \varrho(x, A_i)$$

(see the proof of Theorem 7.3.2).

2.4. Prove that if a subset A of $\mathbf{R}^n$ is convex, then also int A and cl$A$ are convex.

2.5. Prove Theorem 2.3.11.

2.6. (a) Prove that the limit in the set $\mathcal{E}^n$ of hyperplanes (Definition 2.5.2) is induced by some metric in $\mathcal{E}^n$. (Hint: Notice that if ϕ is the parametric representation of $\mathcal{E}^n$ (formula (2.7)), then the function h defined for $E = \phi(v, t)$ by $h(E) = \Delta((t - 1)v, (t + 1)v)$ is a bijection.)

(b) Show that $\mathcal{X}$ is open (closed) in $\mathcal{E}^n$ if and only if $\phi^{-1}(\mathcal{X})$ is open (closed) in $S^{n-1} \times R_+$.

2.7. Prove that if $E, E_k \in \mathcal{E}^n$ for every $k \in N$, and $f : R^n \to R^n$ is an isometry, then

$$E = \lim E_k \implies f(E) = \lim f(E_k)$$

(Theorem 2.5.3).

2.8. Prove that if $H, E, E_k \in \mathcal{E}^n$ and $H \cap E \neq H \neq H \cap E_k$ for every $k \in N$, then

$$\lim E_k = E \implies \lim(E_k \cap H) = E \cap H.$$

2.9. Check whether the set of hyperplanes parallel to a given one is
   a) closed in $\mathcal{E}^n$,
   b) open in $\mathcal{E}^n$,
   c) dense in $\mathcal{E}^n$.

2.10. Check whether the set of hyperplanes passing through a given point is
   a) nowhere dense in $\mathcal{E}^n$,
   b) closed in $\mathcal{E}^n$,
   c) open in $\mathcal{E}^n$.

2.11. Let E be a hyperplane in R^n. Find compact convex sets A_1, A_2 with

$$\pi_E(A_1) \cap \pi_E(A_2) \neq \pi_E(A_1 \cap A_2).$$

Chapter 3

3.1. Draw a picture that illustrates the proof of 3.3.1.

3.2. Prove the converse theorem to 3.3.6.

3.3. Let $\mathcal{B}_A$ be the family of balls in R^n that contain a bounded set A. Prove that A is closed and convex if and only if $A = \bigcap \mathcal{B}_A$.

3.4. Check whether in 3.4.7 the assumption that A is closed is essential.

3.5. Prove that for every affine automorphism $f : R^n \to R^n$ and every subset A of R^n,

$$f(\mathrm{conv}\, A) = \mathrm{conv}\, f(A).$$

3.6. Prove that for every $A, B \subset \mathbb{R}^n$,

$$\operatorname{conv}(A \cap B) \subset \operatorname{conv} A \cap \operatorname{conv} B \qquad (1)$$

and

$$\operatorname{conv} A \cup \operatorname{conv} B \subset \operatorname{conv}(A \cup B). \qquad (2)$$

Give an example of A, B for which equality holds neither in (1) nor in (2).

3.7. Prove that for every $X \in \mathcal{C}^n$,

$$\bar{b}(\operatorname{conv} X) = \bar{b}(X), \quad \operatorname{diam}(\operatorname{conv} X) = \operatorname{diam} X, \quad d(\operatorname{conv} X) = d(X).$$

3.8. Prove that for any convex subsets A, B of $\mathbb{R}^n$,

$$\operatorname{conv}(A \cup B) = \bigcup_{t \in [0; 1]} (1 - t)A + tB.$$

3.9. Find an example of $A \in \mathcal{K}^n$ such that $b(A, u) = \text{const}$ and A is not a ball
   (a) for $n = 2$,
   (b) for $n = 3$.

3.10. Prove that for every $u \in S^{n-1}$ the function $h(\cdot, u)$ is continuous with respect to the Hausdorff metric.

3.11. Let E be an affine subspace of $\mathbb{R}^n$ with $\dim E \in \{0, \ldots, n - 1\}$. Prove that for every subset A of $\mathbb{R}^n$, if A is symmetric with respect to E, then so is $\operatorname{conv} A$.

Chapter 4

4.1. Prove that the function conv is not induced by any transformation of $\mathbb{R}^n$.

4.2. Complete the proof of Theorem 4.2.7.

4.3. Prove that for any affine subspace E of dimension 1, the function S_E preserves $\mathcal{K}^n$ and $\mathcal{K}_0^n$. (See Definition 4.3.1.)

4.4. Prove that for any affine subspace E of dimension 1, the symmetrization S_E preserves volume.

4.5. Prove that for every line E, the symmetrization S_E is not induced by any transformation of $\mathbb{R}^n$.

4.6. Prove Proposition 4.4.5.

4.7. Prove that for any $f_1, \ldots, f_m \in O(n)$ the map $\frac{1}{m} \sum_{i=1}^{m} f_i$ is an isometry if and only if

$$m = 1 \quad \text{or} \quad f_1 = \cdots = f_m.$$

4.8. (T.Ż.)[1] Let P be a convex polygon in $\mathbb{R}^2$ with k vertices ($k \geq 3$). Evaluate the maximal and minimal numbers of vertices of $S_H(P)$ for a hyperplane H.

4.9. Generalize Example 4.2.12 to arbitrary n.

4.10. Generalize Example 4.2.15 to arbitrary n.

Chapter 5

5.1. Prove that $\kappa_n > 1$ for every $n \in \mathbb{N}$. (Hint: Apply the Bieberbach Theorem 5.4.1.)

Chapter 6

6.1. Give an example of a set of simplices that is not a complex, though its union is a geometric polyhedron.

6.2. Give an example of two different triangulations of a polyhedron.

6.3. Let $P \in \mathcal{P}_0^n$. Prove that every simplex of any triangulation of P is a face of some n-dimensional simplex in this triangulation.

6.4. Give examples that illustrate the first part of the proof of Theorem 6.2.4.

6.5. Prove that for any affine subspace E of $\mathbb{R}^n$ and any convex polytope P, if $P \cap E \neq \emptyset$, then $P \cap E$ is a convex polytope.

6.6. Prove that $\mathcal{P}^n$ is an affine invariant (compare 6.2.6).

6.7. Prove that Minkowski addition does not decrease dimension of convex polytopes:

$$\dim(P + Q) \geq \max\{\dim P, \dim Q\}.$$

Check whether the same is true for arbitrary $A, B \neq \emptyset$.

6.8. Prove that in 6.2.9 the sets on the right-hand side of the equality have pairwise disjoint interiors.

---

[1] "(T.Ż.)" means that the exercise was suggested by Tomasz Żukowski.

6.9. Complete and illustrate the proof of Theorem 6.4.4.

6.10. Prove that if n-dimensional convex polytopes P_1, P_2 in $\mathbf{R}^n$ have convex union and disjoint interiors, then $\dim(P_1 \cap P_2) = n - 1$.

6.11. Prove Corollary 6.4.5.

6.12. Prove Proposition 6.5.2. (Hint: Notice that it suffices to prove it for $n = 2$.)

6.13. Prove that there exists a cylindric polytope in $\mathbf{R}^n$ for $n > 3$ that is not equivalent by dissection to a cube. (Hint: See remarks following Theorem 6.4.2.)

6.14. Prove Theorem 6.1.4. To this end, prove first that if a complex $\mathcal{T}_0$ is a subdivision of a triangulation $\mathcal{T}$ (that is, every simplex $S \in \mathcal{T}$ is a union of some simplices in $\mathcal{T}_0$), then

$$\chi(\mathcal{T}_0) = \chi(\mathcal{T}).$$

6.15. Applying 6.1.5 and 6.1.6, prove that $\chi(A) = 1$ for every convex body A in $\mathbf{R}^n$. (Hint: Every convex body in $\mathbf{R}^n$ is homeomorphic to B^n.)

6.16. (T.Ż.) An $A \in \mathcal{K}^n$ is said to be *Minkowski decomposable* if there exist $A_1, A_2 \in \mathcal{K}^n$ such that $A = A_1 + A_2$ and A_2 is not a homothet of A_1 (i.e., there is no $\lambda > 0$ and $v \in \mathbf{R}^n$ with $A_2 = \lambda A_1 + v$). Otherwise, A is said to be *Minkowski indecomposable*.

(a) Prove that every triangle is Minkowski indecomposable (compare [64] Theorem 3.2.11.)

(b) Prove that every convex polygon is the Minkowski sum of some triangles and segments.

6.17. (T.Ż.) Prove that mean width $\bar{b}$ and perimeter l are additive functions on $\mathcal{P}^2$: for every $P_1, P_2 \in \mathcal{P}^2$,

$$\bar{b}(A_1 + A_2) = \bar{b}(A_1) + \bar{b}(A_2), \quad l(A_1 + A_2) = l(A_1) + l(A_2).$$

(Compare Corollary 9.1.2, for $n = 3$).

6.18. (T.Ż.) Apply 6.16 and 6.17 to calculate the mean width of the following subsets of $\mathbf{R}^2$:

(a) segment,
(b) triangle,
(c) parallelogram,
(d) hexagon with pairwise parallel and congruent opposite sides.

6.19. (T.Ż.) Prove that $\bar{b}(P) = \frac{1}{\pi} l(P)$ for every $P \in \mathcal{P}^2$.

6.20. Let P_1 and P_2 be polyhedra contained in affine subspaces E_1 and E_2, respectively, and let $E_1 \cap E_2$ be a singleton. Prove that

$$\mathcal{F}^{(0)}(P_1 + P_2) = \mathcal{F}^{(0)}(P_1) + \mathcal{F}^{(0)}(P_2).$$

(Let us recall that $\mathcal{F}^{(0)}(P)$ is the set of vertices of the polytope P.)

Chapter 7

7.1. Prove that mean width is a valuation, while the minimal width, diameter, and the functional r_0 are not (compare 7.1.6).

7.2. Prove that if a functional $\Phi_0 : \mathcal{P}_n \to \mathbb{R}$ is invariant under an isometry g of $\mathbb{R}^n$, then also $\underline{\Phi}$ and $\overline{\Phi}$ defined by 7.1.9 are invariant under g.

7.3. Prove that our definition of measure of outer normal angle, $\gamma(P, F)$ (Definition 7.2.1), agrees with that given by Schneider ([64], p. 100).

7.4. Prove Theorem 7.2.7 on the monotonicity of $V_{n-1}|\mathcal{P}^n$ for polytopes of dimension less than n.

7.5. Apply Theorem 7.2.10, which concerns the class $\mathcal{P}^n$, to prove that for every $k \in \{0, \dots, n\}$ the functional $V_k : \mathcal{K}^n \to \mathcal{K}^n$ is increasing.

Chapter 8

8.1. Prove the Hadwiger Theorem 8.1.5 for $n = 2$.

8.2. Prove Theorem 8.2.2 modifying the proof of Theorem 8.1.5.

8.3. Let $\Phi = \sum_{i=0}^{n} \alpha_i V_i$. Prove that if Φ is increasing, then $\alpha_i \geq 0$ for $i = 0, \dots, n$.

Chapter 9

9.1. Calculate the mean width of the sets $A_1 = \Delta((0, 0), (1, 1), (0, 1))$ and $A_2 = I^2$ in $\mathbb{R}^2$ and of the sets $A_1 \times \{0\}$ and $A_2 \times \{0\}$ in $\mathbb{R}^3$:
   (a) applying Definition 2.2.8,
   (b) applying Theorem 9.1.1.
   (Compare Exercise 6.18.)

9.2. Prove Proposition 9.2.2.

9.3. Prove that $\mu\{\phi(v, 0) \mid v \in S^{n-1}\} = 0$ (compare Definition 9.2.1).

9.4. Prove the Crofton formulae for $n = 3$ (Theorem 9.2.7).

9.5. According to traditional notation (compare [29]), for any $A \in \mathcal{K}^2$, $l(A)$ is the perimeter of A if int $A \neq \emptyset$, and double length of A if int $A = \emptyset$; $f(A)$ is the area of A. Thus,

$$l(A) = 2V_1(A), \quad f(A) = V_2(A).$$

In view of Proposition 9.2.6 for $n = 2$, there exist α, β such that

$$f(A) = \alpha \int_{\mathcal{E}_A} l(A \cap E)d\mu(E) \quad \text{and} \quad l(A) = \beta \int_{\mathcal{E}_A} d\mu(E) = \beta\mu(\mathcal{E}_A).$$

Calculate α and β.

Exercises to Part II

Chapter 10

10.1. Let A be the regular triangle in $\mathbb{R}^2$ with vertices a, b, c and with sides of length 1. Calculate the curvature measures $\Phi_i(A, X)$ for $i = 0, 1, 2$
 (a) if $X = \{a\}$,
 (b) if $X = \Delta(a, b)$.

10.2. R. Schneider in [64] (p. 70 and 77) defines the spherical image $\sigma(A, X)$ as follows:
$$\sigma(A, X) := \bigcup_{x \in A \cap X} \sigma(A, x),$$
where
$$\sigma(A, x) = \{u \in S^{n-1} | x \in H(A, u)\}$$
for $x \in A \cap X$. Prove that this definition is equivalent to 10.1.8.

10.3. Prove Theorem 10.1.9.

10.4. Prove that for any $P \in \mathcal{P}^n$ and $F \in \mathcal{F}^0(P)$,
$$\gamma_0(P, F) = \gamma(P, F)$$
(compare Definition 7.2.1 and 10.1.10).

10.5. Verify conditions (i) and (ii) in Theorem 10.1.12.

10.6. Prove that the set of weakly continuous, locally defined, and invariant valuations from $\mathcal{K}^n$ into the set of Borel measures on $\mathbf{R}^n$ is closed under addition and multiplication by nonnegative scalars.

10.7. Prove that if $\Phi_{n-1}(A_1, X) = \Phi_{n-1}(A_2, X)$ for every $X \in \mathcal{B}(\mathbf{R}^n)$, then $A_1 = A_2$.

10.8. Explain why Theorem 10.1.18 gives a partial solution of Problem 10.1.17.

10.9. Prove Proposition 10.3.2.

10.10. Let $A \in \mathcal{K}^n$. Prove that A is strictly convex if and only if $\operatorname{bd} A$ does not contain any segment.

10.11. Verify the formulae for $U_\varepsilon(A, X_j)$, $j = 1, 2$, in Example 101..20.

Chapter 11

11.1. Give example of a set A with

$$\operatorname{reach} A \neq \inf\{\operatorname{reach}(A, a) \mid a \in A\}.$$

11.2. Find $\operatorname{reach} A$
   (a) for $A = S^1$ in $\mathbf{R}^2$,
   (b) for $A = \{(x_1, x_2) \in \mathbf{R}^2 \mid x_1^2 = x_2^2\}$.

11.3. Prove Proposition 11.2.4.

11.4. Let $a = (1, 0), b = (0, 1)$, and $A = (\Delta(-a, a) + \Delta(-b, b)) \setminus \operatorname{int} B^2$. Find $\chi(A)$
   (a) proving that A is homeomorphic to a polyhedron and applying 6.1.3 and 6.1.6,
   (b) proving that A is homeomorphic to a member of the family $\mathcal{U}^2$ and applying (11.10) and 11.2.7.

Chapter 12

12.1. Give and example of a subset A of $\mathbf{R}^n$ such that the set of symmetry centers, $C_0(A)$,
   (a) is countable,
   (b) has the cardinality continuum.

12.2. Prove (by induction) formula (12.2).

12.3. Give examples of selectors for the family of triangles in $\mathbf{R}^2$.

12.4. Prove Theorem 12.2.6.

12.5. Describe the relationship between the notion of center of gravity of a finite set (formula (12.5)) and the notion of convex combination (Definition 3.1.1).

12.6. Prove 12.3.7.

12.7. Prove that if $f : \mathbf{R}^2 \to \mathbf{R}^2$ is a similarity and points $p_1, \ldots, p_k, q_1, \ldots, q_l$ satisfy (12.16) for a set A and a point x_0, then $f(p_1), \ldots, f(p_k), f(q_1), \ldots, f(q_l)$ satisfy this condition for the set $f(A)$ and the point $f(x_0)$.

12.8. Prove that the Steiner point map $s : \mathcal{K}^n \to \mathbf{R}^n$ is equivariant under similarities and continuous.

12.9. Find the Steiner point of the polygon $\mathrm{conv}\{a, b, c, d\}$ for
    (a) $a = (0, 0)$, $b = (4, 0)$, $c = (1, 1)$, $d = (3, 1)$,
    (b) $a = (-2, 0)$, $b = (2, 0)$, $c = (0, 1)$, $d = (0, -4)$.

12.10. Prove that for every $A \in \mathcal{K}_0^n$ the function $A \ni x \mapsto A \cap \sigma_x(A) \in \mathcal{K}^n$ is continuous.

12.11. Let $A \in \mathcal{K}_0^n$. For every $x \in A$, let $A_x := A \cap \sigma_x(A)$ (compare (12.18)). Prove that the family $(A_x)_{x \in A}$ is concave, i.e., for every $t \in [0, 1]$,

$$A_{(1-t)x_1 + tx_2} \supset (1 - t)A_{x_1} + tA_{x_2}.$$

12.12. Prove that (under the notation of Exercise 12.11)

$$A_x = A_y \Longrightarrow x = y.$$

12.13. Prove that the pseudocenter of a convex body belongs to its interior: $p(A) \in \mathrm{int}\, A$ for every $A \in \mathcal{K}_0^n$.

12.14. Prove Proposition 12.6.3.

12.15. Prove Proposition 12.6.6.

12.16. Prove that for any subgroup G of $O(n)$ and any convex body A the set of G-pseudocenters $P_G(A)$ is convex.

12.17. Prove that for every $A \in \mathcal{K}_0^n$ and every subgroup G of $O(n)$, the family $(A_{x,G})_{x \in A}$ is concave (compare Exercise 12.11 and formula (12.22)).

12.18. Complete the proof of Theorem 12.7.5:
    (a) prove that $H \cap C$ has the properties required,
    (b) prove that $B_0 \supset C$.

12.19. Let A be a regular m-gon in $\mathbf{R}^2$. Prove that for m odd the set of quasi-centers $Q(A)$ is infinite, while for m even it is a singleton.

12.20. Prove that the Chebyshev point map $\check{c}$ is equivariant under similarities.

12.21. Let $A \in \mathcal{K}^n$ and $x \in \mathbf{R}^n$. Prove that $\varrho_H(A, \{x\}) = R_A(x)$, the radius of the smallest ball with center x containing A.

Chapter 13

13.1. Find the polar lines of $p = (\frac{1}{2}, \frac{1}{2})$ and $q = (1, 2)$ with respect to S^1
    (a) analytically: applying Definition 13.1.2,
    (b) geometrically: applying Theorem 13.1.3.

13.2. Prove 13.2.3 (b).

13.3. Generalize the notion of combinatorial duality on arbitrary convex bodies and prove that Theorem 13.4.3 remains valid for every $A \in \mathcal{K}^n_{00}$.

13.4. Find combinatorial dualities for the pairs of polytopes in 13.4.2.

13.5. Prove (13.12).

13.6. Prove that for every convex body A in $\mathbf{R}^n$ symmetric with respect to 0 and every $x \in \operatorname{int} A \setminus \{0\}$,
$$(A - x)^* \neq A^* - x.$$

13.7. Find a convex body A in $\mathbf{R}^n$ and a point $x \in \operatorname{int} A$ such that
    (a) $V_n((A - x)^*) = V_n(A^*)$;
    (b) $V_n((A - x)^*) \neq V_n(A^*)$.

13.8. Prove that
$$c_{\lambda_n}(A) = 0 \iff s_0(A^*) = 0.$$
(Compare (13.14).)

13.9. Prove that if $a \in S^{n-1} \cap \operatorname{bd}(A^*)$, then $B(a)$ is a support hyperplane of A^* at a.

13.10. Prove that for every $A \subset \mathbf{R}^n$,
$$(\operatorname{cl}A)^* = (\operatorname{conv}A)^* = A^*.$$

13.11. Find conditions sufficient for a subset A of $\mathbf{R}^n$ to satisfy
$$(\operatorname{bd}A)^* = A^*.$$

Exercises to Part III

Chapter 14

14.1. Prove that for every subset A of R^n the set $\ker A$ is closed in R^n and convex.

14.2. Choose several subsets of R^n and find their kernels.

14.3. For any $A \subset R^n$, let the *extended kernel of A* (in symbols $\operatorname{exker} A$) consist of the points of R^n at which A is star-shaped (see 14.1.2).
   (a) Find the extended kernels of chosen sets.
   (b) Show that $\ker A \subset \operatorname{exker} A$ for every $A \subset R^n$ and there exists A with $\ker A = \emptyset \neq \operatorname{exker} A$.

14.4. Prove 14.1.7.

14.5. Prove 14.1.11.

14.6. Prove 14.2.2. (Hint: Show that $\operatorname{cl} \operatorname{int} A = A$ for every $A \in \mathcal{K}_0^n$.)

14.7. Justify Example 14.2.6.

14.8. Give an example of a Hausdorff convergent sequence $(A_k)_{k \in \mathbb{N}}$ such that $A_k \in \mathcal{S}^n$ for every k but $\lim_H A_k \notin \mathcal{S}^n$.

14.9. Prove that for any compact star set B,

$$\delta_{st}(\{a\}, B) = \sup_{y \in \ker B} \delta(\{0\}, B - y) + \sup_{y \in \ker B} \|a - y\|.$$

14.10. Prove 14.4.3.

14.11. Prove 14.4.4.

14.12. Prove that if a star body A in $\mathbf{R}^n$ is not convex, then there exist two points $x, y \in \mathrm{bd}A$ with $\mathrm{relint}\Delta(x, y) \subset \mathbf{R}^n \setminus A$.

14.13. Let E be an affine subspace of $\mathbf{R}^n$ with $\dim E \in \{0, \ldots, n-1\}$. Prove that for every subset A of $\mathbf{R}^n$, if A is symmetric with respect to E, then so is $\ker A$.

14.14. Find a star body A in $\mathbf{R}^2$ with $\mathrm{int}A$ connected and $\ker A$ being
(a) a singleton;
(b) a segment.

Chapter 15

15.1. Prove that if $A \in \mathcal{K}^n_{00}$ and $\mathrm{pos}u \cap \mathrm{bd}A \subset A(u)$, then $\mathrm{bd}A$ is of the class C^1 (of course, the set on the left-hand side of the inclusion is a singleton).

Chapter 16

16.1. Prove that the selector $r_\varphi : \mathcal{K}^n_0 \to \mathbf{R}$ is equivariant under isometries.

16.2. Prove that the selector $\tilde{r}_\varphi : \mathcal{S}^n \to \mathbf{R}$ is equivariant under isometries.

16.3. Prove that the selector $\tilde{r}_\varphi$ is not continuous with respect to the Hausdorff metric.

References

[1] A.D. Aleksandrov, Zur Theorie der gemischten Volumina von konvexen Körpern I: Verallgemeinerung einiger Begriffe der Theorie der konvexen Körper (in Russian), *Mat. Sbornik, N.S.*, 2 (1937), 947–972.

[2] I. Bárány, On the minimal ring containing the boundary of a convex body, *Acta Sci. Math.* 52 (1988), 93–100.

[3] F. Barthe, M. Fradelizi, B. Maurey, A Short Solution to the Busemann–Petty Problem, *Positivity* 3 (1999), 95–100.

[4] P. Billingsley, *Probability and Measure*, John Wiley & Sons, 1979.

[5] W. Blaschke, *Kreis und Kugel*, 1949.

[6] A. Bogdewicz, M. Moszyńska, Čebyšev sets in the space of convex bodies, to appear in *Rend. Circ. Mat. Palermo* (2005).

[7] V. Boltianskii, *Hilbert's Third Problem*, John Wiley & Sons, 1978.

[8] T. Bonnesen, Über das isoperimetrische Defizit ebener Figuren, *Math. Ann.*, 91 (1924), 252–268.

[9] T. Bonnesen, *Les Problèmes des Isopérimètres et des Isépiphanes*, Gauthier-Villars, 1929.

[10] T. Bonnesen, W. Fenchel, *Theorie der konvexen Körper*, Chelsea Publishing Company, 1948.

[11] K. Borsuk, *Multidimensional Analytic Geometry*, Polish Scientific Publishers, 1969.

[12] K. Borsuk, *Theory of Retracts*, PWN, 1967.

[13] J.J. Charatonik, W.J. Charatonik, Inducible mappings between hyperspaces, *Bull. Pol. Acad. Sci.* 46(1) (1998), 5–9.

[14] R. Engelking, *Theory of dimensions, Finite and Infinite*, Heldermann Verlag, 1995.

[15] R. Engelking, *General Topology*, Heldermann Verlag, 1989.

[16] K.J. Falconer, *The Geometry of Fractal Sets*, Cambridge University Press, 1985.

[17] I. Fáry, L. Rédei, Der zentralsymmetrische Kern und die zentralsymmetrische Hülle von konvexen Körpern, *Math. Ann.*, 122 (1950), 205–220.

[18] H. Federer, Curvature measures, *Trans. AMS* 93 (1959), 418–481.

[19] W. Fenchel, B. Jessen, Mengenfunktionen und konvexe Körper, *Danske Vid. Selsk. Mat.-Fys. Medd.*, 16(3) (1938), 1–31.

[20] R.J. Gardner, *Geometric Tomography*, Cambridge University Press, 1995.

[21] R.J. Gardner, A. Volčič, Tomography of convex and star bodies, *Advances in Math.* 108 (1994), 367–399.

[22] R.J. Gardner, A. Koldobsky, T. Schlumprecht, An analytic solution to the Busemann–Petty problem on sections of convex bodies, *Annals of Math.*, 149 (1999), 691–703.

[23] P. Goodey, W. Weil, Intersection Bodies and Ellipsoids, *Mathematika* 42 (1995), 295–304.

[24] H. Groemer, On the Euler Characteristic in Spaces with a Separability Property, *Math. Ann.* 211 (1974), 315–321.

[25] P.M. Gruber, The space of compact subsets of E^d, *Geom. Dedicata* 9 (1980), 87–90.

[26] P. Gruber, G. Lettl, Isometries of the space of convex bodies in Euclidean space, *Bull. London Math. Soc.* 12 (1980), 455–462.

[27] B. Grünbaum, Measures of symmetry for convex sets, *Proc. Symposia Pure Math., Vol. VII, Convexity*; Amer. Math. Soc. (1963), 233–270.

[28] B. Grünbaum, *Convex Polytopes*, John Wiley & Sons, 1967.

[29] H. Hadwiger, *Altes und Neues uber Konvexe Körper*, Birkhäuser, 1955.

[30] H. Hadwiger, *Vorlesungen über Inhalt, Oberfläche und Isoperimetrie*, Springer, 1957.

[31] I. Herburt, On convex hulls of star sets, *Bull. Pol. Acad. Sci.*, 49 (4) (2001), 433–440.

[32] I. Herburt, On the Lipschitz continuity of the centre of the minimal ring, *Rend. Circ. Mat. Palermo*, Serie II, Suppl. 70 (2002), 385–393.

[33] I. Herburt, Convexity of dual volumes for convex bodies, submitted for publication.

[34] I. Herburt, M. Moszyńska, On metric products, *Coll. Math.* 62 (1991), 121–133.

[35] I. Herburt, M. Moszyńska, Z. Peradzyński, Remarks on radial centres of convex bodies, *Analysis and Geometry*, 8(2) (2005), 157–172.

[36] J.G. Hocking, G.S. Young, *Topology*, Addison-Wesley Publ. Comp., London 1961.

[37] A. Illanes, S.B. Nadler, Jr., *Hyperspaces*, Marcel Dekker, Inc., 1999.

[38] D. Klain, A short proof of Hadwiger's characterization theorem, *Mathematika*, 42 (1995), 329–339.

[39] D. Klain, G.-C.Rota, *Introduction to Geometric Probability*, Cambridge University Press, 1997.

[40] M. Kordos, *Streifzüge durch die Mathematik-Geschichte*, Ernst Klett Verlag, Stuttgart 1999.

[41] K. Leichtweiss, *Konvexe Mengen*, Springer, 1980.

[42] E. Lutwak, Dual mixed volumes, *Pacific J. Math.* 58 (1975), 531–538.

[43] E. Lutwak, Intersection Bodies and Dual Mixed Volumes, *Advances in Math.* 71(2) (1988), 232–261.

[44] P. McMullen, *The Polytope Algebra*, Advances in Math. 78 (1989), 76–130.

[45] P. McMullen, R. Schneider, Valuations on convex bodies; in *Convexity and Its Applications*, (ed. by P.M. Gruber and J.M. Wills), Birkhäuser, 1983, 170–247.

[46] M. Moszyńska, Remarks on the minimal rings of convex bodies, *Studia Sci. Math. Hung.* 35 (1999), 1–20.

[47] M. Moszyńska, Quotient star bodies, intersection bodies, and star duality, *J. Math. Anal. Appl.* 232 (1999), 45–60.

[48] M. Moszyńska, Looking for selectors of star bodies, *Geometriae Dedicata* 81 (2000), 131–147.

[49] M. Moszyńska, K. Przesławski, Santaló points associated with radial densities, *Rend. Circ. Mat. Palermo*, (2) Suppl. 50 (1997), 273–288.

[50] M. Moszyńska, J. Święcicka, *Geometry and Linear Algebra* (in Polish), Polish Scientific Publishers, 1987.

[51] M. Moszyńska, T. Żukowski, Duality of convex bodies, *Geometriae Dedicata* 58 (1995), 161–173.

[52] M. Moszyńska, T. Żukowski, On G-pseudo-centres of convex bodies, *Glasnik Mat.* 33(53) (1998), 251–265.

[53] M. Moszyńska, T. Żukowski, On G-pseudo-centres of convex bodies (II), *Rend. Circ. Mat. Palermo*, Serie II, Suppl. 70 (2002), 177–190.

[54] T.S. Motzkin, Sur quelques propriétés caractéristiques des ensembles convexes, *Atti Real. Accad. Naz. Lincei, Rend. Cl. Sci. Fis., Mat., Natur.*, Serie VI,21, (1935), 562–667.

[55] C.M. Petty, Projection bodies; in: *Proceedings, Coll. Convexity, Copenhagen, 1965*, Kobenhavns Univ. Mat. Inst., (1967), 234–241.

[56] K. Przesławski, Linear and Lipschitz continuous selectors for the family of convex sets in Euclidean vector spaces, *Bull. Pol. Acad. Sci.* 33 (1985), 31–33.

[57] C.H. Sah, *Hilbert's Third Prroblem: Scissors Congruence*, Pitman, 1979.

[58] A. Sard, Linear Approximation, AMS, 1963.

[59] R. Schneider, Zu einem Problem von Shephard über die Projektionen konvexer Körper, *Math. Z.* 101 (1967), 71–82.

[60] R. Schneider, On Steiner points of convex bodies, *Israel J. Math.* 9 (1971), 241–249.

[61] R. Schneider, Isometrien des Raumes der konvexen Körper, *Coll. Math.* 33 (1975), 219–224.

[62] R. Schneider, Curvature Measures of Convex Bodies, *Ann. Mat. Pura Appl.* 116 (1978), 101–134.

[63] R. Schneider, Convex surfaces, curvature and surface area measures; in *Handbook of Convex Geometry* (ed. by P.M. Gruber and J.M. Wills), North-Holland, 1993.

[64] R. Schneider, *Convex Bodies: The Brunn–Minkowski Theory*, Cambridge University Press, 1993.

[65] G.C. Shephard, Shadow systems of convex bodies, *Israel J. Math.* 2 (1964), 229–236.

[66] D.W. Stroock, *A Concise Introduction to the Theory of Integration*, Birkhäuser, 1990.

[67] H. Weyl, On the volume of tubes, *Amer. J. Math.* 61 (1939), 461–472.

[68] B. Zdrodowski, On minimal rings for convex bodies (in Polish), master thesis, 1995, Dept. of Math., Computer Sci., and Mech., Warsaw Univ. (unpublished).

[69] G. Zhang, Centered bodies and dual mixed volumes, *Trans. AMS* 345 (1994), 777–801.

[70] G. Zhang, A positive solution to the Busemann–Petty problem in R^4, *Ann. of Math.* 149 (1999), 535–543.

List of Symbols

Index